DISCRETE MATHEMATICS: AN INTRODUCTION

Edited by
Maria Catherine C. Borres

ArclerPress

Discrete Mathematics: An Introduction

Edited by Maria Catherine C. Borres

Edition 2018

ArclerPress

Published by
Arcler Press LLC
708 3rd Avenue, 6th Floor
New York
NY 10017
United States of America
www.arclerpress.com

ISBN: 978-1-68094-583-6 (Hardback)
Printed in United States of America
Library of Congress Control Number: 2017933169

About the Editor

Maria Catherine C. Borres

Catherine is currently taking up Master of Arts in Education Major in Mathematics in Philippine Normal University – Manila. She is currently working as a Content Developer for Mathematics at the Affordable Private Education Center (APEC Schools).

Table of Contents

List of Abbreviations xi
Preface xiii

1 MATHEMATICAL LOGIC
Subfields and Scope 2
History 2
Early History 3
19th Century 3
20th Century 4
Formal Logical Systems 7
First-Order Logic 7
Other Classical Logics 8
Nonclassical and Modal Logic 8
Algebraic Logic 9
Set Theory 9
Model Theory 10
Recursion Theory 10
Algorithmically Unsolvable Problems 11
Proof Theory and Constructive Mathematics 11
Connections With Computer Science 12
Foundations of Mathematics 12
List of First-Order Theories 13
Preliminaries 14
Pure Identity Theories 14
Unary Relations 15
Equivalence Relations 15
Orders 16
Lattices 17
Graphs 17
Boolean Algebras 18
Groups 19

Rings and Fields 20
Geometry 21
Differential Algebra 21
Addition 22
Arithmetic 23
Second Order Arithmetic 24
Set Theories 24
List of Logic Symbols 25
Basic Logic Symbols 26
Advanced and Rarely used Logical Symbols 28
References 30

2 SET THEORY

History 36
Basic Concepts and Notation 37
Some Ontology 38
Axiomatic Set Theory 39
Applications 40
Areas of Study 41
Objections to Set Theory as a Foundation for Mathematics 43
Category Theory 44
Applications of Categories 45
Functors 48
Natural Transformations 48
Other Concepts 49
Historical Notes 50
Relational Model 51
Topics 54
Interpretation 55
Examples 58
Set-Theoretic Formulation 60
Algorithm to Derive Candidate Keys from Functional Dependencies 63
Relational Algebra 63
Joins and Join-Like Operators 65
Common Extensions 71
Use of Algebraic Properties for Query Optimization 74
Implementations 77
Relational Calculus 77
References 78

3 COMBINATORICS

Introduction 83
Approaches and Subfields of Combinatorics 85
Enumerative Combinatorics 85
Analytic Combinatorics 86
Partition Theory 86
Graph Theory 86
Design Theory 87
Finite Geometry 87
Order Theory 88
Matroid Theory 88
Extremal Combinatorics 88
Probabilistic Combinatorics 89
Algebraic Combinatorics 90
Combinatorics on Words 90
Geometric Combinatorics 91
Topological Combinatorics 91
Arithmetic Combinatorics 92
Infinitary Combinatorics 92
Related Fields 92
Combinatorial Optimization 93
Coding Theory 93
Discrete and Computational Geometry 93
Combinatorics and Dynamical Systems 93
Combinatorics and Physics 93
Combinatorial Game Theory 93
Examples 96
Game Abbreviations 97
Nimbers 98
References 98

4 GRAPH THEORY

Introduction 101
Definitions 102
Graph 102
Applications 103
Graph Drawing 107
Graph-Theoretic Data Structures 107
Problems In Graph Theory 108

Enumeration 108
Subgraphs, induced Subgraphs, and Minors...108
Graph Coloring..109
Subsumption and Unification 109
Route Problems 109
Network Flow..110
Visibility Problems..110
Covering Problems 110
Decomposition Problems...110
Graph Classes 110
References..111

5 PROBABILITY THEORY AND DISTRIBUTION

Introduction 113
History 114
Treatment..114
Motivation 114
Discrete Probability Distributions 115
Continuous Probability Distributions...116
Measure-Theoretic Probability Theory 117
Classical Probability Distributions 118
Convergence of Random Variables...118
Weak Convergence..118
Convergence in Probability 119
Law of Large Numbers 119
Central Limit Theorem..120
Probability Distribution...120
Terminology..122
Discrete Probability Distribution..124
Delta-function representation 126
Indicator-function representation..126
Continuous Probability Distribution 126
Some Properties..127
Kolmogorov Definition 128
Random Number Generation...128
Applications..128
Common Probability Distributions..128
Beta Distribution 131
Characterization...131
Properties 134

Geometric mean 138
Measures of Statistical Dispersion 144
Mean absolute difference 150
Parameter Estimation 178
Four Unknown Parameters 179
Four Unknown Parameters 187
Related Distributions 195
References 198

6 ABSTRACT ALGEBRA: CONCEPTS AND APPLICATIONS

Introduction 199
Early Group Theory 201
Modern Algebra 202
Basic Concepts 203
Algebraic Structure 203
Examples 204
Two Sets with Operations 206
Hybrid Structures 207
Universal Algebra 208
Applications 208
References 209

7 DISCRETE GEOMETRY

Introduction 211
Topics in Discrete Geometry 212
Polyhedra and Polytopes 212
Packings, Coverings and Tilings 213
Structural Rigidity and Flexibility 213
Incidence Structures 214
Oriented Matroids 215
Geometric Graph Theory 215
Simplicial Complexes 215
Topological Combinatorics 215
Lattices and Discrete Groups 216
Digital Geometry 216
Discrete Differential Geometry 216
References 217

8 GAME THEORY

Introduction 219
Game Types 222
General and Applied Uses 230
In popular Culture 235
References 236

Index 243

List of Abbreviations

AD	Axiom of determinacy
CLT	central limit theorem
CGT	Combinatorial game theory
DCF	Differentially closed fields
GCLT	Generalized Central Limit Theorem
MK	Morse–Kelley set theory
NBG	Neumann–Bernays–Gödel set theory
RM	Relational model
ZF	Zermelo–Fraenkel set theory
ZFC	Zermelo–Fraenkel set theory

Preface

The world of mathematics can be divided roughly into two realms: the continuous and the discrete. Continuous mathematics studies concepts that are infinite in scope, where one object can blend smoothly into the next. The real-number system lies at the core of continuous mathematics and - just like the watch - between any two real numbers, there is an infinity of real numbers. Continuous mathematics provides excellent models and tools for analyzing real-world phenomena that change smoothly over time, including the motion of planets around the sun or the flow of blood through the body. Discrete mathematics provides excellent models and tools for analyzing real-world phenomena that change abruptly and that lie clearly in one state or another. Discrete mathematics is the tool of choice in a host of applications, from computers to telephone call routing and from personnel assignments to genetics. Discrete mathematics is the branch of mathematics dealing with objects that can assume only distinct, separated values. The term "discrete mathematics" is therefore used in contrast with "continuous mathematics," which is the branch of mathematics dealing with objects that can vary smoothly. Whereas discrete objects can often be characterized by integers, continuous objects require real numbers. It is increasingly being applied in the practical fields of mathematics and computer science. It is a very good tool for improving reasoning and problem-solving capabilities. The study of how discrete objects combine with one another and the probabilities of various outcomes is known as combinatorics. Other fields of mathematics that are considered to be part of discrete mathematics include graph theory and the theory of computation. Topics in number theory such as congruences and recurrence relations are also considered part of discrete mathematics. The study of topics in discrete mathematics usually includes the study of algorithms, their implementations, and efficiencies. Discrete mathematics is the mathematical language of computer science, and as such, its importance has increased dramatically in recent decades.

Content Coverage

Chapter One aims to present the concept of mathematical logic that explores the applications of formal logic to mathematics. It bears close connections to metamathematics, the foundations of mathematics, and theoretical computer science.

Chapter Two is set theory that deals exclusively with sets, so the only sets under consideration are those whose members are also sets. The language of set theory that can be used in the definitions of nearly all mathematical objects.

Chapter Three is concerned the study of finite or countable discrete structures. Aspects of combinatorics include counting the structures of a given kind and size.

Chapter Four focuses on graph theory that deals with the study of points and lines. In particular, it involves the ways in which sets of points, called vertices, can be connected by lines or arcs, called edges. Graphs in this context differ from the more familiar coordinate plots that portray mathematical relations and functions.

Chapter Five is concerned with probability, the analysis of random phenomena. The central objects of probability theory are random variables, stochastic processes, and events: mathematical abstractions of non-deterministic events or measured quantities that may either be single occurrences or evolve over time in an apparently random fashion.

Chapter Six focuses on the concepts and applications of abstract that deal with abstract algebraic structures rather than the usual number systems. The most important of these structures are groups, rings, and fields.

Chapter Seven is concerned with properties of finitely generated geometric objects such as polytopes and polyhedra, triangulations and polyhedral complexes, configurations of lines and, more generally, hyperplanes in Euclidean and other spaces, the theory of rigid and flexible frameworks, tilings and packings.

Chapter Eight presents the study of mathematical models of conflict and cooperation between intelligent rational decision-makers.

CHAPTER 1

Mathematical Logic

Chapter Contents

- **Subfields and Scope**
- **History**
- **Formal Logical Systems**
- **Set Theory**
- **Model Theory**
- **Recursion Theory**
- **Proof Theory and Constructive Mathematics**
- **Connections with Computer Science**
- **Foundations of Mathematics**
- **List of First-Order Theories**
- **List of Logic Symbols**

The rules of mathematical logic specify methods of reasoning mathematical statements. Greek philosopher, Aristotle, was the pioneer of logical reasoning. Logical reasoning provides the theoretical base for many areas of mathematics and consequently computer science. It has many practical applications in computer science like design of computing machines, artificial intelligence, definition of data structures for programming languages etc.

Starting with the basics of set theory, induction and computability, it covers propositional and first-order logic — their syntax, reasoning systems and semantics. Soundness and completeness results for Hilbert's and Gentzen's systems are presented, along with simple decidability arguments. The general applicability of various concepts and techniques is demonstrated by highlighting their consistent reuse in different contexts.

Mathematical logic is a subfield of mathematics exploring the applications of formal logic to mathematics. It bears close connections to metamathematics, the foundations of mathematics, and theoretical computer science. The unifying themes in mathematical logic include the study of the expressive power of formal systems and the deductive power of formal proof systems.

Mathematical logic is often divided into the fields of set theory, model theory, recursion theory, and proof theory. These areas share basic results on logic, particularly first-order logic, and definability. In computer science (particularly in the ACM Classification) mathematical logic encompasses additional topics not detailed in this article; see Logic in computer science for those.

Since its inception, mathematical logic has both contributed to, and has been motivated by, the study of foundations of mathematics. This study began in the late 19th century with the development of axiomatic frameworks for geometry, arithmetic, and analysis. In the early 20th century it was shaped by David Hilbert's program to prove the consistency of foundational theories. Results of Kurt Gödel, Gerhard Gentzen, and others provided partial resolution to the program, and clarified the issues involved in proving consistency. Work in set theory showed that almost all ordinary mathematics can be formalized in terms of sets, although there are some theorems that cannot be proven in common axiom systems for set theory. Contemporary work in the foundations of mathematics often focuses on establishing which parts of mathematics can be formalized in particular formal systems (as in reverse mathematics) rather than trying to find theories in which all of mathematics can be developed.

SUBFIELDS AND SCOPE

The *Handbook of Mathematical Logic* (Barwise 1989) makes a rough division of contemporary mathematical logic into four areas:

- set theory
- model theory
- recursion theory, and
- proof theory and constructive mathematics (considered as parts of a single area).

Each area has a distinct focus, although many techniques and results are shared among multiple areas. The borderlines amongst these fields, and the lines separating mathematical logic and other fields of mathematics, are not always sharp. Gödel's incompleteness theorem marks not only a milestone in recursion theory and proof theory, but has also led to Löb's theorem in modal logic. The method of forcing is employed in set theory, model theory, and recursion theory, as well as in the study of intuitionistic mathematics.

The mathematical field of category theory uses many formal axiomatic methods, and includes the study of categorical logic, but category theory is not ordinarily considered a subfield of mathematical logic. Because of its applicability in diverse fields of mathematics, mathematicians including Saunders Mac Lane have proposed category theory as a foundational system for mathematics, independent of set theory. These foundations use toposes, which resemble generalized models of set theory that may employ classical or nonclassical logic.

HISTORY

Mathematical logic emerged in the mid-19th century as a subfield of mathematics independent of the traditional study of logic (Ferreirós 2001, p. 443). Before this emergence, logic was studied with rhetoric, through the syllogism, and with philosophy. The first half of the 20th century saw an explosion of fundamental results, accompanied by vigorous debate over the foundations of mathematics.

Early History

Further information: History of logic

Theories of logic were developed in many cultures in history, including China, India, Greece and the Islamic world. In 18th-century Europe, attempts to treat the operations of formal logic in a symbolic or algebraic way had been made by philosophical mathematicians including Leibniz and Lambert, but their labors remained isolated and little known.

19th Century

In the middle of the nineteenth century, George Boole and then Augustus De Morgan presented systematic mathematical treatments of logic. Their work, building on work by algebraists such as George Peacock, extended the traditional Aristotelian doctrine of logic into a sufficient framework for the study of foundations of mathematics (Katz 1998, p. 686).

Charles Sanders Peirce built upon the work of Boole to develop a logical system for relations and quantifiers, which he published in several papers from 1870 to 1885. Gottlob Frege presented an independent development of logic with quantifiers in his *Begriffsschrift*, published in 1879, a work generally considered as marking a turning point in the history of logic. Frege's work remained obscure, however, until Bertrand Russell began to promote it near the turn of the century. The two-dimensional notation Frege developed was never widely adopted and is unused in contemporary texts.

From 1890 to 1905, Ernst Schröder published *Vorlesungen über die Algebra der Logik* in three volumes. This work summarized and extended the work of Boole, De Morgan, and Peirce, and was a comprehensive reference to symbolic logic as it was understood at the end of the 19th century.

Foundational Theories

Concerns that mathematics had not been built on a proper foundation led to the development of axiomatic systems for fundamental areas of mathematics such as arithmetic, analysis, and geometry.

In logic, the term *arithmetic* refers to the theory of the natural numbers. Giuseppe Peano (1889) published a set of axioms for arithmetic that came to bear his name (Peano axioms), using a variation of the logical system of Boole and Schröder but adding quantifiers. Peano was unaware of Frege's work at the time. Around the same time Richard Dedekind showed that the natural numbers are uniquely characterized by their induction properties. Dedekind (1888) proposed a different characterization, which lacked the formal logical character of Peano's axioms. Dedekind's work, however, proved theorems inaccessible in Peano's system, including the uniqueness of the set of natural numbers (up to isomorphism) and the recursive definitions of addition and multiplication from the successor function and mathematical induction.

In the mid-19th century, flaws in Euclid's axioms for geometry became known (Katz 1998, p. 774). In addition to the independence of the parallel postulate, established by Nikolai Lobachevsky in 1826 (Lobachevsky 1840), mathematicians discovered that certain theorems taken for granted by Euclid were not in fact provable from his axioms. Among these is the theorem that a line contains at least two points, or that circles of the same radius whose centers are separated by that radius must intersect. Hilbert (1899) developed a complete set of axioms for geometry, building on previous work by Pasch (1882). The success in axiomatizing geometry motivated Hilbert to seek complete axiomatizations of other areas of mathematics, such as the natural numbers and the real line. This would prove to be a major area of research in the first half of the 20th century.

The 19th century saw great advances in the theory of real analysis, including theories of convergence of functions and Fourier series. Mathematicians such as Karl Weierstrass began to construct functions that stretched intuition, such as nowhere-differentiable continuous functions. Previous conceptions of a function as a rule for computation, or a smooth graph, were no longer adequate. Weierstrass began to advocate the arithmetization of analysis, which sought to axiomatize analysis using properties of the natural numbers. The modern (ε, δ)-definition of limit and continuous functions was already developed by Bolzano in 1817 (Felscher 2000), but remained relatively unknown. Cauchy in 1821 defined continuity in terms of infinitesimals (see Cours d'Analyse, page 34). In 1858, Dedekind proposed a definition of the real numbers in terms of Dedekind cuts of rational numbers (Dedekind 1872), a definition still employed in contemporary texts.

Georg Cantor developed the fundamental concepts of infinite set theory. His early results developed the theory of cardinality and proved that the reals and the natural numbers have different cardinalities (Cantor 1874). Over the next twenty years, Cantor developed a theory of transfinite numbers in a series of publications. In 1891, he published a new proof of the uncountability of the real numbers that introduced the diagonal argument, and used this method to prove Cantor's theorem that no set can have the same cardinality as its powerset. Cantor believed that every set could be well-ordered, but was unable to produce a proof for this result, leaving it as an open problem in 1895 (Katz 1998, p. 807).

20th Century

In the early decades of the 20th century, the main areas of study were set theory and formal logic. The discovery of paradoxes in informal set theory caused some to wonder whether mathematics itself is inconsistent, and to look for proofs of consistency.

In 1900, Hilbert posed a famous list of 23 problems for the next century. The first two of these were to resolve the continuum hypothesis and prove the consistency of elementary arithmetic, respectively; the tenth was to produce a method that could decide whether a multivariate polynomial equation over the integers has a solution. Subsequent work to resolve these problems shaped the direction of mathematical logic, as did the effort to resolve Hilbert's *Entscheidungsproblem*, posed in 1928. This problem asked for a procedure that would decide, given a formalized mathematical statement, whether the statement is true or false.

Set Theory and Paradoxes

Ernst Zermelo (1904) gave a proof that every set could be well-ordered, a result Georg Cantor had been unable to obtain. To achieve the proof, Zermelo introduced the axiom of choice, which drew heated debate and research among mathematicians and the pioneers of set theory. The immediate criticism of the method led Zermelo to publish a second exposition of his result, directly addressing criticisms of his proof (Zermelo 1908a). This paper led to the general acceptance of the axiom of choice in the mathematics community.

Skepticism about the axiom of choice was reinforced by recently discovered paradoxes in naive set theory. Cesare Burali-Forti (1897) was the first to state a paradox: the Burali-Forti paradox shows that the collection of all ordinal numbers cannot form a set. Very soon thereafter, Bertrand Russell discovered Russell's paradox in 1901, and Jules Richard (1905) discovered Richard's paradox.

Zermelo (1908b) provided the first set of axioms for set theory. These axioms, together with the additional axiom of replacement proposed by Abraham Fraenkel, are now called Zermelo–Fraenkel set theory (ZF). Zermelo's axioms incorporated the principle of limitation of size to avoid Russell's paradox.

In 1910, the first volume of *Principia Mathematica* by Russell and Alfred North Whitehead was published. This seminal work developed the theory of functions and cardinality in a completely formal framework of type theory, which Russell and Whitehead developed in an effort to avoid the paradoxes. *Principia Mathematica* is considered one of the most influential works of the 20th century, although the framework of type theory did not prove popular as a foundational theory for mathematics (Ferreirós 2001, p. 445).

Fraenkel (1922) proved that the axiom of choice cannot be proved from the remaining axioms of Zermelo's set theory with urelements. Later work by Paul Cohen (1966) showed that the addition of urelements is not needed, and the axiom of choice is unprovable in ZF. Cohen's proof developed the method of forcing, which is now an important tool for establishing independence results in set theory.

Symbolic Logic

Leopold Löwenheim (1915) and Thoralf Skolem (1920) obtained the Löwenheim–Skolem theorem, which says that first-order logic cannot control the cardinalities of infinite structures. Skolem realized that this theorem would apply to first-order formalizations of set theory, and that it implies any such formalization has a countable model. This counterintuitive fact became known as Skolem's paradox.

In his doctoral thesis, Kurt Gödel (1929) proved the completeness theorem, which establishes a correspondence between syntax and semantics in first-order logic. Gödel used the completeness theorem to prove the compactness theorem, demonstrating the finitary nature of first-order logical consequence. These results helped establish first-order logic as the dominant logic used by mathematicians.

In 1931, Gödel published *On Formally Undecidable Propositions of Principia Mathematica and Related Systems*, which proved the incompleteness (in a different meaning of the word) of all sufficiently strong, effective first-order theories. This result, known as Gödel's incompleteness theorem, establishes severe limitations on axiomatic foundations for mathematics, striking a strong blow to Hilbert's program. It showed the impossibility of providing a consistency proof of arithmetic within any formal theory of arithmetic. Hilbert, however, did not acknowledge the importance of the incompleteness theorem for some time.

Gödel's theorem shows that a consistency proof of any sufficiently strong, effective axiom system cannot be obtained in the system itself, if the system is consistent, nor in any weaker system. This leaves open the possibility of consistency proofs that cannot be formalized within the system they consider. Gentzen (1936) proved the consistency of arithmetic using a finitistic system together with a principle of transfinite induction. Gentzen's result introduced the ideas of cut elimination and proof-theoretic ordinals, which became key tools in proof theory. Gödel (1958) gave a different consistency proof, which reduces the consistency of classical arithmetic to that of intuitionistic arithmetic in higher types.

Beginnings of the Other Branches

Alfred Tarski developed the basics of model theory.

Beginning in 1935, a group of prominent mathematicians collaborated under the pseudonym Nicolas Bourbaki to publish a series of encyclopedic mathematics texts. These texts, written in an austere and axiomatic style, emphasized rigorous presentation and set-theoretic foundations. Terminology coined by these texts, such as the words *bijection*, *injection*, and *surjection*, and the set-theoretic foundations the texts employed, were widely adopted throughout mathematics.

The study of computability came to be known as recursion theory, because early formalizations by Gödel and Kleene relied on recursive definitions of functions. When these definitions were shown equivalent to Turing's formalization involving Turing machines, it became clear that a new concept – the computable function – had been discovered, and that this definition was robust enough to admit numerous independent characterizations. In his work on the incompleteness theorems in 1931, Gödel lacked a rigorous concept of an effective formal system; he immediately realized that the new definitions of computability could be used for this purpose, allowing him to state the incompleteness theorems in generality that could only be implied in the original paper.

Numerous results in recursion theory were obtained in the 1940s by Stephen Cole Kleene and Emil Leon Post. Kleene (1943) introduced the concepts of relative computability, foreshadowed by Turing (1939), and the arithmetical hierarchy. Kleene later generalized recursion theory to higher-order functionals. Kleene and Kreisel studied formal versions of intuitionistic mathematics, particularly in the context of proof theory.

FORMAL LOGICAL SYSTEMS

At its core, mathematical logic deals with mathematical concepts expressed using formal logical systems. These systems, though they differ in many details, share the common property of considering only expressions in a fixed formal language. The systems of propositional logic and first-order logic are the most widely studied today, because of their applicability to foundations of mathematics and because of their desirable proof-theoretic properties. Stronger classical logics such as second-order logic or infinitary logic are also studied, along with nonclassical logics such as intuitionistic logic.

First-Order Logic

First-order logic is a particular formal system of logic. Its syntax involves only finite expressions as well-formed formulas, while its semantics are characterized by the limitation of all quantifiers to a fixed domain of discourse.

Early results from formal logic established limitations of first-order logic. The Löwenheim–Skolem theorem (1919) showed that if a set of sentences in a countable first-order language has an infinite model then it has at least one model of each infinite cardinality. This shows that it is impossible for a set of first-order axioms to characterize the natural numbers, the real numbers, or any other infinite structure up to isomorphism. As the goal of early foundational studies was to produce axiomatic theories for all parts of mathematics, this limitation was particularly stark.

Gödel's completeness theorem (Gödel 1929) established the equivalence between semantic and syntactic definitions of logical consequence in first-order logic. It shows that if a particular sentence is true in every model that satisfies a particular set of axioms, then there must be a finite deduction of the sentence from the axioms. The compactness theorem first appeared as a lemma in Gödel's proof of the completeness theorem, and it took many years before logicians grasped its significance and began to apply it routinely. It says that a set of sentences has a model if and only if every finite subset has a model, or in other words that an inconsistent set of formulas must have a finite inconsistent subset. The completeness and compactness theorems allow for sophisticated analysis of logical consequence in first-order logic and the development of model theory, and they are a key reason for the prominence of first-order logic in mathematics.

Gödel's incompleteness theorems (Gödel 1931) establish additional limits on first-order axiomatizations. The first incompleteness theorem states that for any consistent, effectively given (defined below) logical system that is capable of interpreting arithmetic, there exists a statement that is true but not provable within that logical system (and which indeed may fail in some non-standard models of arithmetic which may be consistent with the logical system). For example, in every logical system capable of expressing the Peano axioms, the Gödel sentence holds for the natural numbers but cannot be proved.

Here a logical system is said to be effectively given if it is possible to decide, given any formula in the language of the system, whether the formula is an axiom, and one which can express the Peano axioms is called "sufficiently strong." When applied to first-order logic, the first incompleteness theorem implies that any sufficiently strong, consistent, effective first-order theory has models that are not elementarily equivalent, a stronger limitation than the one established by the Löwenheim–Skolem theorem. The **second incompleteness theorem** states that no sufficiently strong, consistent, effective axiom system for arithmetic can prove its own consistency, which has been interpreted to show that Hilbert's program cannot be completed.

Other Classical Logics

Many logics besides first-order logic are studied. These include infinitary logics, which allow for formulas to provide an infinite amount of information, and higher-order logics, which include a portion of set theory directly in their semantics.

The most well studied infinitary logic is $L_{\omega_1,\omega}$. In this logic, quantifiers may only be nested to finite depths, as in first-order logic, but formulas may have finite or countably infinite conjunctions and disjunctions within them. Thus, for example, it is possible to say that an object is a whole number using a formula of $L_{\omega_1,\omega}$ such as

$$(x = 0) \vee (x = 1) \vee (x = 2) \vee \cdots.$$

Higher-order logics allow for quantification not only of elements of the domain of discourse, but subsets of the domain of discourse, sets of such subsets, and other objects of higher type. The semantics are defined so that, rather than having a separate domain for each higher-type quantifier to range over, the quantifiers instead range over all objects of the appropriate type. The logics studied before the development of first-order logic, for example Frege's logic, had similar set-theoretic aspects. Although higher-order logics are more expressive, allowing complete axiomatizations of structures such as the natural numbers, they do not satisfy analogues of the completeness and compactness theorems from first-order logic, and are thus less amenable to proof-theoretic analysis.

Another type of logics are fixed-point logics that allow inductive definitions, like one writes for primitive recursive functions.

One can formally define an extension of first-order logic — a notion which encompasses all logics in this section because they behave like first-order logic in certain fundamental ways, but does not encompass all logics in general, e.g. it does not encompass intuitionistic, modal or fuzzy logic. Lindström's theorem implies that the only extension of first-order logic satisfying both the compactness theorem and the Downward Löwenheim–Skolem theorem is first-order logic.

Nonclassical and Modal Logic

Modal logics include additional modal operators, such as an operator which states that a particular formula is not only true, but necessarily true. Although modal logic is not often used to axiomatize mathematics, it has been used to study the properties of first-order provability (Solovay 1976) and set-theoretic forcing (Hamkins and Löwe 2007).

Intuitionistic logic was developed by Heyting to study Brouwer's program of intuitionism, in which Brouwer himself avoided formalization. Intuitionistic logic specifically does not include the law of the excluded middle, which states that each sentence is either true or its negation is true. Kleene's work with the proof theory of intuitionistic logic showed that constructive information can be recovered from intuitionistic proofs. For example, any provably total function in intuitionistic arithmetic is computable; this is not true in classical theories of arithmetic such as Peano arithmetic.

Algebraic Logic

Algebraic logic uses the methods of abstract algebra to study the semantics of formal logics. A fundamental example is the use of Boolean algebras to represent truth values in classical propositional logic, and the use of Heyting algebras to represent truth values in intuitionistic propositional logic. Stronger logics, such as first-order logic and higher-order logic, are studied using more complicated algebraic structures such as cylindric algebras.

SET THEORY

Set theory is the study of sets, which are abstract collections of objects. Many of the basic notions, such as ordinal and cardinal numbers, were developed informally by Cantor before formal axiomatizations of set theory were developed. The first such axiomatization, due to Zermelo (1908b), was extended slightly to become Zermelo–Fraenkel set theory (ZF), which is now the most widely used foundational theory for mathematics.

Other formalizations of set theory have been proposed, including von Neumann–Bernays–Gödel set theory (NBG), Morse–Kelley set theory (MK), and New Foundations (NF). Of these, ZF, NBG, and MK are similar in describing a cumulative hierarchy of sets. New Foundations takes a different approach; it allows objects such as the set of all sets at the cost of restrictions on its set-existence axioms. The system of Kripke–Platek set theory is closely related to generalized recursion theory.

Two famous statements in set theory are the axiom of choice and the continuum hypothesis. The axiom of choice, first stated by Zermelo (1904), was proved independent of ZF by Fraenkel (1922), but has come to be widely accepted by mathematicians. It states that given a collection of nonempty sets there is a single set C that contains exactly one element from each set in the collection. The set C is said to "choose" one element from each set in the collection. While the ability to make such a choice is considered obvious by some, since each set in the collection is nonempty, the lack of a general, concrete rule by which the choice can be made renders the axiom nonconstructive. Stefan Banach and Alfred Tarski (1924) showed that the axiom of choice can be used to decompose a solid ball into a finite number of pieces which can then be rearranged, with no scaling, to make two solid balls of the original size. This theorem, known as the Banach–Tarski paradox, is one of many counterintuitive results of the axiom of choice.

The continuum hypothesis, first proposed as a conjecture by Cantor, was listed by David Hilbert as one of his 23 problems in 1900. Gödel showed that the continuum hypothesis cannot be disproven from the axioms of Zermelo–Fraenkel set theory (with or without the axiom of choice), by developing the constructible universe of set theory in which the continuum hypothesis must hold. In 1963, Paul Cohen showed that the continuum hypothesis cannot be proven from the axioms of Zermelo–Fraenkel set theory (Cohen 1966). This independence result did not completely settle Hilbert's question, however, as it is possible that new axioms for set theory could resolve the hypothesis. Recent work along these lines has been conducted by W. Hugh Woodin, although its importance is not yet clear (Woodin 2001).

Contemporary research in set theory includes the study of large cardinals and determinacy. Large cardinals are cardinal numbers with particular properties so strong that the existence of such cardinals cannot be proved in ZFC. The existence of the smallest large cardinal typically studied, an inaccessible cardinal, already implies the consistency of ZFC. Despite the fact that large cardinals have extremely high cardinality, their existence has many ramifications for the structure of the real line. *Determinacy* refers to the possible existence of winning strategies for certain two-player games (the games are said to be *determined*). The existence of these strategies implies structural properties of the real line and other Polish spaces.

MODEL THEORY

Model theory studies the models of various formal theories. Here a theory is a set of formulas in a particular formal logic and signature, while a model is a structure that gives a concrete interpretation of the theory. Model theory is closely related to universal algebra and algebraic geometry, although the methods of model theory focus more on logical considerations than those fields.

The set of all models of a particular theory is called an elementary class; classical model theory seeks to determine the properties of models in a particular elementary class, or determine whether certain classes of structures form elementary classes.

The method of quantifier elimination can be used to show that definable sets in particular theories cannot be too complicated. Tarski (1948) established quantifier elimination for real-closed fields, a result which also shows the theory of the field of real numbers is decidable. (He also noted that his methods were equally applicable to algebraically closed fields of arbitrary characteristic.) A modern subfield developing from this is concerned with o-minimal structures.

Morley's categoricity theorem, proved by Michael D. Morley (1965), states that if a first-order theory in a countable language is categorical in some uncountable cardinality, i.e. all models of this cardinality are isomorphic, then it is categorical in all uncountable cardinalities.

A trivial consequence of the continuum hypothesis is that a complete theory with less than continuum many nonisomorphic countable models can have only countably many. Vaught's conjecture, named after Robert Lawson Vaught, says that this is true even independently of the continuum hypothesis. Many special cases of this conjecture have been established.

RECURSION THEORY

Recursion theory, also called **computability theory**, studies the properties of computable functions and the Turing degrees, which divide the uncomputable functions into sets that have the same level of uncomputability. Recursion theory also includes the study of generalized computability and definability. Recursion theory grew from the work of Alonzo Church and Alan Turing in the 1930s, which was greatly extended by Kleene and Post in the 1940s.

Classical recursion theory focuses on the computability of functions from the natural numbers to the natural numbers. The fundamental results establish a robust, canonical class of computable functions with numerous independent, equivalent characterizations using Turing machines, λ calculus, and other systems. More advanced results concern the structure of the Turing degrees and the lattice of recursively enumerable sets.

Generalized recursion theory extends the ideas of recursion theory to computations that are no longer necessarily finite. It includes the study of computability in higher types as well as areas such as hyperarithmetical theory and α-recursion theory.

Contemporary research in recursion theory includes the study of applications such as algorithmic randomness, computable model theory, and reverse mathematics, as well as new results in pure recursion theory.

Algorithmically Unsolvable Problems

An important subfield of recursion theory studies algorithmic unsolvability; a decision problem or function problem is **algorithmically unsolvable** if there is no possible computable algorithm that returns the correct answer for all legal inputs to the problem. The first results about unsolvability, obtained independently by Church and Turing in 1936, showed that the Entscheidungsproblem is algorithmically unsolvable. Turing proved this by establishing the unsolvability of the halting problem, a result with far-ranging implications in both recursion theory and computer science.

There are many known examples of undecidable problems from ordinary mathematics. The word problem for groups was proved algorithmically unsolvable by Pyotr Novikov in 1955 and independently by W. Boone in 1959. The busy beaver problem, developed by Tibor Radó in 1962, is another well-known example.

Hilbert's tenth problem asked for an algorithm to determine whether a multivariate polynomial equation with integer coefficients has a solution in the integers. Partial progress was made by Julia Robinson, Martin Davis and Hilary Putnam. The algorithmic unsolvability of the problem was proved by Yuri Matiyasevich in 1970 (Davis 1973).

PROOF THEORY AND CONSTRUCTIVE MATHEMATICS

Proof theory is the study of formal proofs in various logical deduction systems. These proofs are represented as formal mathematical objects, facilitating their analysis by mathematical techniques. Several deduction systems are commonly considered, including Hilbert-style deduction systems, systems of natural deduction, and the sequent calculus developed by Gentzen.

The study of **constructive mathematics**, in the context of mathematical logic, includes the study of systems in non-classical logic such as intuitionistic logic, as well as the study of predicative systems. An early proponent of predicativism was Hermann Weyl, who showed it is possible to develop a large part of real analysis using only predicative methods (Weyl 1918).

Because proofs are entirely finitary, whereas truth in a structure is not, it is common for work in constructive mathematics to emphasize provability. The relationship between provability in classical (or nonconstructive) systems and provability in intuitionistic (or constructive, respectively) systems is of particular interest. Results such as the Gödel–Gentzen negative translation show that it is possible to embed (or *translate*) classical logic into intuitionistic logic, allowing some properties about intuitionistic proofs to be transferred back to classical proofs.

Recent developments in proof theory include the study of proof mining by Ulrich Kohlenbach and the study of proof-theoretic ordinals by Michael Rathjen.

CONNECTIONS WITH COMPUTER SCIENCE

The study of computability theory in computer science is closely related to the study of computability in mathematical logic. There is a difference of emphasis, however. Computer scientists often focus on concrete programming languages and feasible computability, while researchers in mathematical logic often focus on computability as a theoretical concept and on noncomputability.

The theory of semantics of programming languages is related to model theory, as is program verification (in particular, model checking). The Curry–Howard isomorphism between proofs and programs relates to proof theory, especially intuitionistic logic. Formal calculi such as the lambda calculus and combinatory logic are now studied as idealized programming languages. Computer science also contributes to mathematics by developing techniques for the automatic checking or even finding of proofs, such as automated theorem proving and logic programming.

Descriptive complexity theory relates logics to computational complexity. The first significant result in this area, Fagin's theorem (1974) established that NP is precisely the set of languages expressible by sentences of existential second-order logic.

FOUNDATIONS OF MATHEMATICS

In the 19th century, mathematicians became aware of logical gaps and inconsistencies in their field. It was shown that Euclid's axioms for geometry, which had been taught for centuries as an example of the axiomatic method, were incomplete. The use of infinitesimals, and the very definition of function, came into question in analysis, as pathological examples such as Weierstrass' nowhere-differentiable continuous function were discovered.

Cantor's study of arbitrary infinite sets also drew criticism. Leopold Kronecker famously stated "God made the integers; all else is the work of man," endorsing a return to the study of finite, concrete objects in mathematics. Although Kronecker's argument was carried forward by constructivists in the 20th century, the mathematical community as a whole rejected them. David Hilbert argued in favor of the study of the infinite, saying "No one shall expel us from the Paradise that Cantor has created."

Mathematicians began to search for axiom systems that could be used to formalize large parts of mathematics. In addition to removing ambiguity from previously naive terms such as function, it was hoped that this axiomatization would allow for consistency proofs. In the 19th century, the main method of proving the consistency of a set of axioms was to provide a model for it. Thus, for example, non-Euclidean geometry can be proved consistent by defining *point* to mean a point on a fixed sphere and *line* to mean a great circle on the sphere. The resulting structure, a model of elliptic geometry, satisfies the axioms of plane geometry except the parallel postulate.

With the development of formal logic, Hilbert asked whether it would be possible to prove that an axiom system is consistent by analyzing the structure of possible proofs in the system, and showing through this analysis that it is impossible to prove a contradiction. This idea led to the study of proof theory. Moreover, Hilbert proposed that the analysis should be entirely concrete, using the term *finitary* to refer to the methods he would allow but not precisely defining them. This project, known as Hilbert's program, was seriously affected by Gödel's incompleteness theorems, which show that the consistency of formal theories of arithmetic cannot be established using methods formalizable in those theories. Gentzen showed that it is possible to produce a proof of the consistency of arithmetic in a finitary system augmented with axioms of transfinite induction, and the techniques he developed to do so were seminal in proof theory.

A second thread in the history of foundations of mathematics involves nonclassical logics and constructive mathematics. The study of constructive mathematics includes many different programs with various definitions of *constructive*. At the most accommodating end, proofs in ZF set theory that do not use the axiom of choice are called constructive by many mathematicians. More limited versions of constructivism limit themselves to natural numbers, number-theoretic functions, and sets of natural numbers (which can be used to represent real numbers, facilitating the study of mathematical analysis). A common idea is that a concrete means of computing the values of the function must be known before the function itself can be said to exist. In the early 20th century, Luitzen Egbertus Jan Brouwer founded intuitionism as a philosophy of mathematics. This philosophy, poorly understood at first, stated that in order for a mathematical statement to be true to a mathematician, that person must be able to *intuit* the statement, to not only believe its truth but understand the reason for its truth. A consequence of this definition of truth was the rejection of the law of the excluded middle, for there are statements that, according to Brouwer, could not be claimed to be true while their negations also could not be claimed true. Brouwer's philosophy was influential, and the cause of bitter disputes among prominent mathematicians. Later, Kleene and Kreisel would study formalized versions of intuitionistic logic (Brouwer rejected formalization, and presented his work in unformalized natural language). With the advent of the BHK interpretation, Kripke models or computability logic, intuitionism became easier to reconcile with classical mathematics.

LIST OF FIRST-ORDER THEORIES

In mathematical logic, a first-order theory is given by a set of axioms in some language. This entry lists some of the more common examples used in model theory and some of their properties.

Preliminaries

For every natural mathematical structure there is a signature σ listing the constants, functions, and relations of the theory together with their valences, so that the object is naturally a σ-structure. Given a signature σ there is a unique first-order language L_σ that can be used to capture the first-order expressible facts about the σ-structure.

There are two common ways to specify theories:

- List or describe a set of sentences in the language L_σ, called the **axioms** of the theory.
- Give a set of σ-structures, and define a theory to be the set of sentences in L_σ holding in all these models. For example, the «theory of finite fields» consists of all sentences in the language of fields that are true in all finite fields.

An L_σ theory may:

- be consistent: no proof of contradiction exists;
- be satisfiable: there exists a σ-structure for which the sentences of the theory are all true (by the completeness theorem, satisfiability is equivalent to consistency);
- be complete: for any statement, either it or its negation is provable;
- have quantifier elimination;
- eliminate imaginaries;
- be finitely axiomatizable;
- be decidable: There is an algorithm to decide which statements are provable;
- be recursively axiomatizable;
- be Model complete or sub-model complete;
- be κ-categorical: All models of cardinality κ are isomorphic;
- be Stable or unstable.
- be ω-stable (same as totally transcendental for countable theories).
- be superstable
- have an atomic model
- have a prime model
- have a saturated model

Pure Identity Theories

The signature of the pure identity theory is empty, with no functions, constants, or relations.

Pure identity theory has no (non-logical) axioms. It is decidable.

One of the few interesting properties that can be stated in the language of pure identity theory is that of being infinite. This is given by an infinite set of axioms stating there are at least 2 elements, there are at least 3 elements, and so on:

- $\exists x_1 \exists x_2 \neg x_1 = x_2$, $\quad \exists x_1 \exists x_2 \exists x_3 \neg x_1 = x_2 \wedge \neg x_1 = x_3 \wedge \neg x_2 = x_3, \ldots$

These axioms define the **theory of an infinite set.**

The opposite property of being finite cannot be stated in first-order logic for any theory that has arbitrarily large finite models: in fact any such theory has infinite models by the compactness theorem. In general if a property can be stated by a finite number of sentences of first-order logic then the opposite property can also be stated in first-order logic, but if a property needs an infinite number of sentences then its opposite property cannot be stated in first-order logic.

Any statement of pure identity theory is equivalent to either $\sigma(N)$ or to $\neg\sigma(N)$ for some finite subset N of the non-negative integers, where $\sigma(N)$ is the statement that the number of elements is in N. It is even possible to describe all possible theories in this language as follows. Any theory is either the theory of all sets of cardinality in N for some *finite* subset N of the non-negative integers, or the theory of all sets whose cardinality is not in N, for some *finite or infinite* subset N of the non-negative integers. (There are no theories whose models are exactly sets of cardinality N if N is an infinite subset of the integers.) The complete theories are the theories of sets of cardinality n for some finite n, and the theory of infinite sets.

One special case of this is the **inconsistent theory** defined by the axiom $\exists x \neg x = x$. It is a perfectly good theory with many good properties: it is complete, decidable, finitely axiomatizable, and so on. The only problem is that it has no models at all. By Gödel's completeness theorem, it is the only theory (for any given language) with no models. It is not the same as the theory of the empty set (in versions of first-order logic that allow a model to be empty): the theory of the empty set has exactly one model, which has no elements.

Unary Relations

A set of unary relations P_i for i in some set I is called independent if for every two disjoint finite subsets A and B of I there is some element x such that $P_i(x)$ is true for i in A and false for i in B. Independence can be expressed by a set of first-order statements.

The theory of a countable number of independent unary relations is complete, but has no atomic models. It is also an example of a theory that is superstable but not totally transcendental.

Equivalence Relations

The signature of equivalence relations has one binary infix relation symbol ~, no constants, and no functions. Equivalence relations satisfy the axioms:

- **Reflexivity** $\forall x\ x \sim x$;
- **Symmetry** $\forall x\ \forall y\ x \sim y \rightarrow y \sim x$;
- **Transitivity**: $\forall x\ \forall y\ \forall z\ (x \sim y \land y \sim z) \rightarrow x \sim z$.

Some first order properties of equivalence relations are:

- ~ has an infinite number of equivalence classes;
- ~ has exactly n equivalence classes (for any fixed positive integer n);

- All equivalence classes are infinite;
- All equivalence classes have size exactly n (for any fixed positive integer n).

The theory of an equivalence relation with exactly 2 infinite equivalence classes is an easy example of a theory which is ω-categorical but not categorical for any larger cardinal.

The equivalence relation ~ should not be confused with the identity symbol '=': if $x=y$ then $x \sim y$, but the converse is not necessarily true. Theories of equivalence relations are not all that difficult or interesting, but often give easy examples or counterexamples for various statements.

The following constructions are sometimes used to produce examples of theories with certain spectra; in fact by applying them to a small number of explicit theories T one gets examples of complete countable theories with all possible uncountable spectra. If T is a theory in some language, we define a new theory 2^T by adding a new binary relation to the language, and adding axioms stating that it is an equivalence relation, such that there are an infinite number of equivalence classes all of which are models of T. It is possible to iterate this construction transfinitely: given an ordinal α, define a new theory by adding an equivalence relation E for each $\beta<\alpha$, together with axioms stating that whenever $\beta<\gamma$ then each E equivalence class is the union of infinitely many E_β equivalence classes, and each E_0 equivalence class is a model of T. Informally, one can visualize models of this theory as infinitely branching trees of height α with models of T attached to all leaves.

Orders

The signature of orders has no constants or functions, and one binary relation symbols $\leq$. (It is of course possible to use $\geq$, $<$ or $>$ instead as the basic relation, with the obvious minor changes to the axioms.) We define $x \geq y$, $x < y$, $x > y$ as abbreviations for $y \leq x$, $x \leq y \land \neg y \leq x$, $y < x$,

Some first-order properties of orders:

- **Transitive**: $\forall x\ \forall y\ \forall z\ x \leq y \land y \leq z \rightarrow x \leq z$
- **Reflexive**: $\forall x\ x \leq x$
- **Antisymmetric**: $\forall x\ \forall y\ x \leq y \land y \leq x \rightarrow x = y$
- **Partial**: Transitive∧Reflexive∧Antisymmetric;
- **Linear** (or **total**): Partial ∧ $\forall x\ \forall y\ x \leq y \lor y \leq x$
- **Dense** $\forall x\ \forall z\ x < z \rightarrow \exists y\ x < y \land y < z$ ("Between any 2 distinct elements there is another element")
- There is a smallest element: $\exists x\ \forall y\ x \leq y$
- There is a largest element: $\exists x\ \forall y\ y \leq x$
- Every element has an immediate successor: $\forall x\ \exists y\ \forall z\ x < z \leftrightarrow y \leq z$

The theory DLO of *dense linear orders without endpoints* (i.e. no smallest or largest element) is complete, ω-categorical, but not categorical for any uncountable cardinal. There are 3 other very similar theories: the theory of dense linear orders with a:

- Smallest but no largest element;
- Largest but no smallest element;
- Largest and smallest element.

Being **well ordered** ("any non-empty subset has a minimal element") is not a first-order property; the usual definition involves quantifying over all *subsets*.

Lattices

Lattices can be considered either as special sorts of partially ordered sets, with a signature consisting of one binary relation symbol ≤, or as algebraic structures with a signature consisting of two binary operations ∧ and ∨. The two approaches can be related by defining $a \leq b$ to mean $a \wedge b = a$.

For two binary operations the axioms for a lattice are:

Commutative laws: $\forall a \forall b\, a \vee b = b \vee a$ $\forall a \forall b\, a \wedge b = b \wedge a$

Associative laws:

$\forall a \forall b \forall c\, a \vee (b \vee c) = (a \vee b) \vee c$ $\forall a \forall b \forall c\, a \wedge (b \wedge c) = (a \wedge b) \wedge c$

Absorption laws: $\forall a \forall b\, a \vee (a \wedge b) = a$ $\forall a \forall b\, a \wedge (a \vee b) = a$

For one relation ≤ the axioms are:

- Axioms stating ≤ is a partial order, as above.
- $\forall a \forall b \exists c\, c \leq a \wedge c \leq b \wedge \forall d\, d \leq a \wedge d \leq b \rightarrow d \leq c$ (existence of c=a∧b)
- $\forall a \forall b \exists c\, a \leq c \wedge b \leq c \wedge \forall d\, a \leq d \wedge b \leq d \rightarrow c \leq d$ (existence of c=a∨b)

First order properties include:

- $\forall x \forall y \forall z\, x \vee (y \wedge z) = (x \vee y) \wedge (x \vee z)$ (distributive lattices)
- $\forall x \forall y \forall z\, x \vee (y \wedge (x \vee z)) = (x \vee y) \wedge (x \vee z)$ (modular lattices)

Heyting algebras can be defined as lattices with certain extra first-order properties.

Completeness is not a first order property of lattices.

Graphs

The signature of graphs has no constants or functions, and one binary relation symbol R, where $R(x,y)$ is read as "there is an edge from x to y".

The axioms for the **theory of graphs** are

- **Symmetric**: $\forall x\ \forall y\ R(x,y) \rightarrow R(y,x)$
- **Anti-reflexive**: $\forall x\ \neg R(x,x)$ ("no loops")

The *theory of random graphs* has the following extra axioms for each positive integer n:

- For any two disjoint finite sets of size n, there is a point joined to all points of the first set and to no points of the second set. (For each fixed n, it is easy to write this statement in the language of graphs.)

The theory of random graphs is ω categorical, complete, and decidable, and its countable model is called the Rado graph. A statement in the language of graphs is true in this theory if and only if the probability that an n-vertex random graph models the statement tends to 1 in the limit as n goes to infinity.

Boolean Algebras

There are several different signatures and conventions used for **Boolean algebras**:

- The signature has 2 constants, 0 and 1, and two binary functions ∧ and ∨ ("and" and "or"), and one unary function ¬ ("not"). This is a little confusing as the functions use the same symbols as the propositional functions of first-order logic.
- In set theory, a common convention is that the language has 2 constants, 0 and 1, and two binary functions · and +, and one unary function −. The three functions have the same interpretation as the functions in the first convention. Unfortunately, this convention clashes badly with the next convention:
- In algebra, the usual convention is that the language has 2 constants, 0 and 1, and two binary functions · and +. The function · has the same meaning as ∧, but $a+b$ means $a \vee b \wedge \neg(a \wedge b)$. The reason for this is that the axioms for a Boolean algebra are then just the axioms for a ring with 1 plus $\forall x\ x^2 = x$. Unfortunately this clashes with the standard convention in set theory given above.

The axioms are:

- The axioms for a distributive lattice (see above)
- $\forall a\ a \wedge \neg a = 0$, $\forall a\ a \vee \neg a = 1$ (properties of negation)
- Some authors add the extra axiom $\neg 0 = 1$, to exclude the trivial algebra with one element.

Tarski proved that the theory of Boolean algebras is decidable.

We write $x \leq y$ as an abbreviation for $x \wedge y = x$, and atom(x) as an abbreviation for $\neg x = 0 \wedge \forall y\ y \leq x \rightarrow y = 0 \vee y = x$, read as "$x$ is an atom", in other words a non-zero element with nothing between it and 0. Here are some first-order properties of Boolean algebras:

- **Atomic**: $\forall x\ x=0 \vee \exists y\ y \leq x \wedge \text{atom}(y)$
- **Atomless**: $\forall x\ \neg\text{atom}(x)$

The theory of **atomless Boolean algebras** is ω-categorical and complete.

For any Boolean algebra B, there are several invariants defined as follows.

- the ideal $I(B)$ consists of elements that are the sum of an atomic and an atomless element (an element with no atoms below it).
- The quotient algebras B^i of B are defined inductively by $B^0=B$, $B^{k+1} = B^k/I(B^k)$.
- The invariant $m(B)$ is the smallest integer such that B^{m+1} is trivial, or ∞ if no such integer exists.
- If $m(B)$ is finite, the invariant $n(B)$ is the number of atoms of $B^{m(B)}$ if this

number is finite, or ∞ if this number is infinite.

- The invariant $l(B)$ is 0 if $B^{m(B)}$ is atomic or if $m(B)$ is ∞, and 1 otherwise.

Then two Boolean algebras are elementarily equivalent if and only if their invariants l, m, and n are the same. In other words, the values of these invariants classify the possible completions of the theory of Boolean algebras. So the possible complete theories are:

- The trivial algebra (if this is allowed; sometimes $0 \neq 1$ is included as an axiom.)
- The theory with $m = \infty$
- The theories with m a natural number, n a natural number or ∞, and $l = 0$ or 1 (with $l = 0$ if $n = 0$).

Groups

The signature of group theory has one constant 1 (the identity), one function of arity 1 (the inverse) whose value on t is denoted by t^{-1}, and one function of arity 2, which is usually omitted from terms. For any integer n. t^n is an abbreviation for the obvious term for the nth power of t.

Groups are defined by the axioms

- *Identity*: $\forall x\ 1x = x \wedge x1 = x$
- *Inverse*: $\forall x\ x^{-1}x = 1 \wedge xx^{-1} = 1$
- *Associative*: $\forall x \forall y \forall z\ (xy)z = x(yz)$

Some properties of groups that can be defined in the first-order language of groups are:

- **Abelian** $\forall x\ \forall y\ xy = yx$.
- **Torsion free** $\forall x\ x^2 = 1 \rightarrow x = 1$, $\forall x\ x^3 = 1 \rightarrow x = 1$, $\forall x\ x^4 = 1 \rightarrow x = 1$, ...
- **Divisible** $\forall x\ \exists y\ y^2 = x$, $\forall x\ \exists y\ y^3 = x$, $\forall x\ \exists y\ y^4 = x$, ...
- **Infinite** (as in identity theory)
- **Exponent** n (for any fixed positive integer n) $\forall x\ x^n = 1$
- Nilpotent of class n (for any fixed positive integer n)
- Solvable of class n (for any fixed positive integer n)

The theory of **abelian groups** is decidable. The theory of **Infinite divisible torsion-free abelian groups** is complete, as is the theory of **Infinite abelian groups of exponent p** (for p prime).

The theory of **finite groups** is the set of first-order statements in the language of groups that are true in all finite groups (there are plenty of infinite models of this theory). It is not completely trivial to find any such statement that is not true for all groups: one example is "given two elements of order 2, either they are conjugate or there is a non-trivial element commuting with both of them".

The properties of being finite, or free, or simple, or torsion are not first-order. More precisely, the first-order theory of all groups with one of these properties has models that do not have this property.

Rings and Fields

The signature of (unital) rings has 2 constants 0 and 1, two binary functions + and ×, and, optionally, one unary negation function −.

Rings Axioms: Addition makes the ring into an abelian group, multiplication is associative and has an identity 1, and multiplication is left and right distributive.

Commutative rings The axioms for rings plus $\forall x\ \forall y\ xy=yx$.

Fields The axioms for commutative rings plus $\forall x\ \neg\ x=0 \rightarrow \exists y\ xy=1$ and $\neg\ 1=0$. Many of the examples given here have only universal, or *algebraic* axioms. The class of structures satisfying such a theory has the property of being closed under substructure. For example, a subset of a group closed under the group actions of multiplication and inverse is again a group. Since the signature of fields does not usually include multiplicative and additive inverse, the axioms for inverses are not universal, and therefore a substructure of a field closed under addition and multiplication is not always a field. This can be remedied by adding unary inverse functions to the language.

For any positive integer n the property that all equations of degree n have a root can be expressed by a single first-order sentence:

- $\forall\ a_1\ \forall\ a_2 ...\ \forall\ a_n\ \exists x\ (...((x+a_1)x\ +a_2)x+...)x+a_n = 0$

Perfect fields The axioms for fields, plus axioms for each prime number p stating that if p 1 = 0 (i.e. the field has characteristic p), then every field element has a pth root.

Algebraically closed fields of characteristic *p* The axioms for fields, plus for every positive n the axiom that all polynomials of degree n have a root, plus axioms fixing the characteristic. The classical examples of complete theories. Categorical in all uncountable cardinals. The theory ACF_p has a *universal domain property*, in the sense that every structure N satisfying the universal axioms of ACF_p is a substructure of a sufficiently large algebraically closed field $M \models ACF_0$, and additionally any two such embeddings $N \rightarrow M$ induce an automorphism of M.

Finite fields. The theory of finite fields is the set of all first-order statements that are true in all finite fields. Significant examples of such statements can, for example, be given by applying the Chevalley–Warning theorem, over the prime fields. The name is a little misleading as the theory has plenty of infinite models. Ax proved that the theory is decidable.

Formally real fields These are fields with the axiom

- For every positive n, the axiom $\forall\ a_1\ \forall\ a_2 ...\ \forall\ a_n\ a_1a_1+a_2a_2+\ ...+a_na_n=0 \rightarrow a_1=0 \vee a_2=0 \vee\ ...\ \vee a_n=0$ (0 is not a non-trivial sum of squares).

Real closed fields Axioms:

- $\forall x\ \exists y\ x=yy \vee x+yy=0$.
- For every odd positive n, the axiom stating that every polynomial of degree n has a root.
- For every positive n, the axiom $\forall\ a_1\ \forall\ a_2 ...\ \forall\ a_n\ a_1a_1+a_2a_2+\ ...+a_na_n=0 \rightarrow a_1=0 \vee a_2=0 \vee\ ...\ \vee a_n=0$ (0 is not a non-trivial sum of squares).

The theory of real closed fields is effective and complete and therefore decidable (the Tarski-Seidenberg theorem). The addition of further function symbols (e.g., the exponential function, the sine function) may change decidability.

***p*-adic fields**: Ax & Kochen (1965) showed that the theory of p-adic fields is decidable and gave a set of axioms for it.

Geometry

Axioms for various systems of geometry usually use a typed language, with the different types corresponding to different geometric objects such as points, lines, circles, planes, and so on. The signature will often consist of binary incidence relations between objects of different types; for example, the relation that a point lies on a line. The signature may have more complicated relations; for example ordered geometry might have a ternary "betweenness" relation for 3 points, which says whether one lies between two others, or a "congruence" relation between 2 pairs of points.

Some examples of axiomatized systems of geometry include ordered geometry, absolute geometry, affine geometry, Euclidean geometry, projective geometry, and hyperbolic geometry. For each of these geometries there are many different and inequivalent systems of axioms for various dimensions. Some of these axiom systems include "completeness" axioms that are not first order.

As a typical example, the axioms for projective geometry use 2 types, points and lines, and a binary incidence relation between points and lines. If point and line variables are indicated by small and capital letter, and a incident to A is written as aA, then one set of axioms is

- $\forall a \forall b\, \neg a = b \rightarrow \exists C\, aC \wedge bC$ (There is a line through any 2 distinct points a,b ...)
- $\forall a \forall b \forall C \forall D\, \neg a = b \wedge aC \wedge bC \wedge aD \wedge bD \rightarrow C = D$ (... which is unique)
- $\forall a \forall b \forall c \forall d \forall e \forall G \forall H\, aH \wedge bH \wedge eH \wedge cG \wedge dG \wedge eG \rightarrow \exists f \exists I \exists J\, aI \wedge cI \wedge fI \wedge bJ \wedge dJ \wedge fJ$ (Veblen's axiom: if ab and cd lie on intersecting lines, then so do ac and bd.)
- $\forall A \exists b \exists c \exists d\, bA \wedge cA \wedge dA \wedge \neg b = c \wedge \neg b = d \wedge \neg c = d$ (Every line has at least 3 points)

Euclid did not state all the axioms for Euclidean geometry explicitly, and the first complete list was given by Hilbert in Hilbert's axioms. This is not a first order axiomatization as one of Hilbert's axioms is a second order completeness axiom. Tarski's axioms are a first order axiomatization of Euclidean geometry. Tarski showed this axiom system is complete and decidable by relating it to the complete and decidable theory of real closed fields.

Differential Algebra

- The theory DF of differential fields.

The signature is that of fields (0, 1, +, −, ×) together with a unary function ∂, the derivation. The axioms are those for fields together with

$$\forall u \forall v\, \partial(uv) = u\,\partial v + v\,\partial u$$
$$\forall u \forall v\, \partial(u+v) = \partial u + \partial v\,.$$

For this theory one can add the condition that the characteristic is p, a prime or zero, to get the theory DF_p of **differential fields of characteristic *p*** (and similarly with the other theories below).

If K is a differential field then the **field of constants** $k = \{u \in K : \partial(u) = 0\}$ The theory of **differentially perfect fields** is the theory of differential fields together with the condition that the field of constants is perfect; in other words, for each prime p it has the axiom:

$$\forall u\, \partial(u) = 0 \wedge p1 = 0 \rightarrow \exists v\, v^p = u$$

(There is little point in demanding that the whole field should be a perfect field, because in non-zero characteristic this implies the differential is 0.) For technical reasons to do with quantifier elimination, it is sometimes more convenient to force the constant field to be perfect by adding a new symbol r to the signature with the axioms

$$\forall u\, \partial(u) = 0 \wedge p1 = 0 \rightarrow r(u)^p = u$$
$$\forall u\, \neg\partial(u) = 0 \rightarrow r(u) = 0.$$

- The theory of **differentially closed fields** (DCF) is the theory of differentially perfect fields with axioms saying that if f and g are differential polynomials and the separant of f is nonzero and $g \neq 0$ and f has order greater than that of g, then there is some x in the field with $f(x)=0$ and $g(x) \neq 0$.

Addition

The **theory of the natural numbers with a successor function** has signature consisting of a constant 0 and a unary function S ("successor": $S(x)$ is interpreted as $x+1$), and has axioms:

- $\forall x\, \neg\, Sx = 0$
- $\forall x \forall y\, Sx = Sy \rightarrow x = y$
- Let $P(x)$ be a first-order formula with a single free variable x. Then the following formula is an axiom:

$(P(0) \wedge \forall x(P(x) \rightarrow P(Sx))) \rightarrow \forall y\, P(y)$.

The last axiom (induction) can be replaced by the axioms

- For each integer $n>0$, the axiom $\forall x$ SSS...Sx $\neq$ x (with n copies of S)
- $\forall x\, \neg\, x = 0 \rightarrow \exists y\, Sy = x$

The theory of the natural numbers with a successor function is complete and decidable, and is κ-categorical for uncountable κ but not for countable κ.

Presburger arithmetic is the theory of the natural numbers under addition, with signature consisting of a constant 0, a unary function S, and a binary function +. It is complete and decidable. The axioms are

- $\forall x \neg Sx = 0$
- $\forall x \forall y\ Sx = Sy \rightarrow x = y$
- $\forall x\ x + 0 = x$
- $\forall x \forall y\ x + Sy = S(x + y)$
- Let $P(x)$ be a first-order formula with a single free variable x. Then the following formula is an axiom:

$(P(0) \wedge \forall x(P(x) \rightarrow P(Sx))) \rightarrow \forall y\ P(y)$.

Arithmetic

Many of the first order theories described above can be extended to complete recursively enumerable consistent theories. This is no longer true for most of the following theories; they can usually encode both multiplication and addition of natural numbers, and this gives them enough power to encode themselves, which implies that Gödel's incompleteness theorem applies and the theories can no longer be both complete and recursively enumerable (unless they are inconsistent).

The signature of a theory of arithmetic has:

- The constant 0;
- The unary function, the successor function, here denoted by prefix S, or by prefix σ or postfix $'$ elsewhere;
- Two binary functions, denoted by infix + and ×, called "addition" and "multiplication."

Some authors take the signature to contain a constant 1 instead of the function S, then define S in the obvious way as $St = 1 + t$.

Robinson arithmetic (also called **Q**). Axioms (1) and (2) govern the distinguished element 0. (3) assures that S is an injection. Axioms (4) and (5) are the standard recursive definition of addition; (6) and (7) do the same for multiplication. Robinson arithmetic can be thought of as Peano arithmetic without induction. **Q** is a weak theory for which Gödel's incompleteness theorem holds. Axioms:

- $\forall x \neg Sx = 0$
- $\forall x \neg x = 0 \rightarrow \exists y\ Sy = x$
- $\forall x \forall y\ Sx = Sy \rightarrow x = y$
- $\forall x\ x + 0 = x$
- $\forall x \forall y\ x + Sy = S(x + y)$
- $\forall x\ x \times 0 = 0$
- $\forall x \forall y\ x \times Sy = (x \times y) + x$.

IΣ_n is first order Peano arithmetic with induction restricted to Σ_n formulas (for n = 0, 1, 2, ...). The theory $I\Sigma_0$ is often denoted by $I\Delta_0$. This is a series of more and more powerful fragments of Peano arithmetic. The case $n = 1$ has about the same strength as **primitive recursive arithmetic** (PRA). **Exponential function arithmetic** (EFA) is $I\Sigma_0$ with an axiom stating that x^y exists for all x and y (with the usual properties).

First order Peano arithmetic, PA. The "standard" theory of arithmetic. The axioms are the axioms of Robinson arithmetic above, together with the axiom scheme of induction:

- {\displaystyle \phi (0)\wedge (\forall x\phi (x)\rightarrow \phi (Sx))\rightarrow (\forall x\phi (x))} for any formula φ in the language of **PA**. φ may contain free variables other than x.

Kurt Gödel's 1931 paper proved that **PA** is incomplete, and has no consistent recursively enumerable completions.

Complete arithmetic (also known as **true arithmetic**) is the theory of the standard model of arithmetic, the natural numbers **N**. It is complete but does not have a recursively enumerable set of axioms.

For the **real numbers**, the situation is slightly different: The case that includes just addition and multiplication cannot encode the integers, and hence Gödel's incompleteness theorem does not apply. Complications arise when adding further function symbols (e.g., exponentiation).

Second Order Arithmetic

Second-order arithmetic can refer to a first order theory (in spite of the name) with two types of variables, thought of as varying over integers and subsets of the integers. (There is also a theory of arithmetic in second order logic that is called second order arithmetic. It has only one model, unlike the corresponding theory in first order logic, which is incomplete.) The signature will typically be the signature 0, S, +, × of arithmetic, together with a membership relation ∈ between integers and subsets (though there are numerous minor variations). The axioms are those of Robinson arithmetic, together with axiom schemes of induction and comprehension.

There are many different subtheories of second order arithmetic that differ in which formulas are allowed in the induction and comprehension schemes. In order of increasing strength, five of the most common systems are

- RCA_0, Recursive Comprehension
- WKL_0, Weak König's lemma
- ACA_0, Arithmetical comprehension
- ATR_0, Arithmetical Transfinite Recursion
- $\Pi^1_1\text{-}\mathsf{CA}_0$, Π^1_1 comprehension

These are defined in detail in the articles on second order arithmetic and reverse mathematics.

Set Theories

The usual signature of set theory has one binary relation ∈, no constants, and no functions. Some of the theories below are "class theories" which have two sorts of object, sets and classes. There are three common ways of handling this in first-order logic:

- Use first-order logic with two types.
- Use ordinary first-order logic, but add a new unary predicate "Set", where "Set(t)" means informally "t is a set".
- Use ordinary first-order logic, and instead of adding a new predicate to the language, treat "Set(t)" as an abbreviation for "$\exists y\ t \in y$"

Some first order set theories include:

- Weak theories lacking powersets:
  - S' (Tarski, Mostowski, and Robinson, 1953); (finitely axiomatizable)
  - General set theory;
  - Kripke-Platek set theory; KP;
  - Pocket set theory
- Mac Lane set theory and elementary topos theory
- Zermelo set theory; Z
- Ackermann set theory;
- Zermelo-Fraenkel set theory; ZF, ZFC;
- Von Neumann-Bernays-Gödel set theory; NBG; (finitely axiomatizable)
- Morse–Kelley set theory; MK;
- Tarski–Grothendieck set theory; TG;
- New Foundations; (finitely axiomatizable)
- Scott-Potter set theory
- Positive set theory

Some extra first order axioms that can be added to one of these (usually ZF) include:

- axiom of choice, axiom of dependent choice
- Generalized continuum hypothesis
- Martin's axiom (usually together with the negation of the continuum hypothesis), Martin's maximum
- ◊ and ♣
- Axiom of constructibility (V=L)
- proper forcing axiom
- analytic determinacy, projective determinacy, Axiom of determinacy
- Many large cardinal axioms

LIST OF LOGIC SYMBOLS

In logic, a set of symbols is commonly used to express logical representation. The following table lists many common symbols together with their name, pronunciation, and the related field of mathematics. Additionally, the third column contains an informal definition, the fourth column gives a short example, the fifth and sixth give the unicode location and name for use in HTML documents. The last column provides the LaTeX symbol.

Outside logic, different symbols have the same meaning, and the same symbol can have different meanings, depending on the context.

Basic Logic Symbols

Symbol	Name Read as Category	Explanation	Examples	Unicode Value (hexdecimal)	HTML Value (decimal)	HTML Entity (named)	LaTeX symbol
⇒ → ⊃	material implication implies; if .. then propositional logic, Heyting algebra	$A \Rightarrow B$ is true only in the case that either A is false or B is true. → may mean the same as ⇒ (the symbol may also indicate the domain and codomain of a function; see table of mathematical symbols). ⊃ may mean the same as ⇒ (the symbol may also mean superset).	$x = 2 \Rightarrow x^2 = 4$ is true, but $x^2 = 4 \Rightarrow x = 2$ is in general false (since x could be −2).	U+21D2 U+2192 U+2283	⇒ → ⊃	⇒ → ⊃	⇒\Rightarrow →\to ⊃\supset ⟹ \implies
⇔ ≡ ↔	material equivalence if and only if; iff; means the same as propositional logic	$A \Leftrightarrow B$ is true only if both A and B are false, or both A and B are true.	$x + 5 = y + 2 \Leftrightarrow x + 3 = y$	U+21D4 U+2261 U+2194	⇔ ≡ ↔	⇔ ≡ ↔	⇔ \Leftrightarrow ≡\equiv ↔ \leftrightarrow ⟺ \iff
¬ ˜ !	negation not propositional logic	The statement ¬A is true if and only if A is false. A slash placed through another operator is the same as "¬" placed in front.	$\neg(\neg A) \Leftrightarrow A$ $x \neq y \Leftrightarrow \neg(x = y)$	U+00AC U+02DC U+0021	¬ ˜ !	¬ ˜ !	¬\lnot or \neg ~\sim
∧ · &	logical conjunction and propositional logic, Boolean algebra	The statement $A \wedge B$ is true if A and B are both true; else it is false.	$n < 4 \wedge n > 2 \Leftrightarrow n = 3$ when n is a natural number.	U+2227 U+00B7 U+0026	∧ · &	∧ · &	∧\wedge or \land &\&[2]

∨ + ∥	logical (inclusive) disjunction or propositional logic, Boolean algebra	The statement $A \lor B$ is true if A or B (or both) are true; if both are false, the statement is false.	$n \geq 4 \ \lor \ n \leq 2 \ \Leftrightarrow n \neq 3$ when n is a natural number.	U+2228 U+002B U+2225	∨ + ∥	∨	∨\lor or \vee ∥\parallel
⊕ ⊻	exclusive disjunction xor propositional logic, Boolean algebra	The statement $A \oplus B$ is true when either A or B, but not both, are true. $A \veebar B$ means the same.	$(\neg A) \oplus A$ is always true, $A \oplus A$ is always false.	U+2295 U+22BB	⊕ ⊻	⊕	⊕\oplus ⊻\veebar
⊤ T 1	Tautology top, verum propositional logic, Boolean algebra	The statement ⊤ is unconditionally true.	$A \Rightarrow \top$ is always true.	U+22A4	⊤		⊤\top
⊥ F 0	Contradiction bottom, falsum, falsity propositional logic, Boolean algebra	The statement ⊥ is unconditionally false. (The symbol ⊥ may also refer to perpendicular lines.)	$\bot \Rightarrow A$ is always true.	U+22A5	⊥	⊥	⊥\bot
∀ ()	universal quantification for all; for any; for each first-order logic	$\forall x$: $P(x)$ or (x) $P(x)$ means $P(x)$ is true for all x.	$\forall n \in \mathbb{N}: n^2 \geq n$.	U+2200	∀	∀	∀\forall
∃	existential quantification there exists first-order logic	$\exists x$: $P(x)$ means there is at least one x such that $P(x)$ is true.	$\exists n \in \mathbb{N}$: n is even.	U+2203	∃	∃	∃\exists
∃!	uniqueness quantification there exists	$\exists! x$: $P(x)$ means there is exactly one x	$\exists! n \in \mathbb{N}: n + 5 = 2n$.	U+2203 U+0021	∃ !		∃!\exists !

	exactly one first-order logic	such that $P(x)$ is true.					
:= ≡ :⇔	definition is defined as everywhere	$x := y$ or $x \equiv y$ means x is defined to be another name for y (but note that ≡ can also mean other things, such as congruence). $P :\Leftrightarrow Q$ means P is defined to be logically equivalent to Q.	$\cosh x := (1/2)(\exp x + \exp(-x))$ $A \text{ XOR } B :\Leftrightarrow (A \vee B) \wedge \neg(A \wedge B)$	U+2254 (U+003A U+003D) U+2261 U+003A U+229C	≔ (: =) ≡ ⊜	≡ ⇔	≔ := ≡\equiv :⇔ :\Leftrightarrow
( )	precedence grouping parentheses, brackets everywhere	Perform the operations inside the parentheses first.	$(8 \div 4) \div 2 = 2 \div 2 = 1$, but $8 \div (4 \div 2) = 8 \div 2 = 4$.	U+0028 U+0029	( )		( ) ( )
⊢	Turnstile provable propositional logic, first-order logic	$x \vdash y$ means y is provable from x (in some specified formal system).	$A \to B \vdash \neg B \to \neg A$	U+22A2	⊢		⊢\vdash
⊨	double turnstile entails propositional logic, first-order logic	$x \vDash y$ means x semantically entails y	$A \to B \vDash \neg B \to \neg A$	U+22A8	⊨		⊨\vDash

Advanced and Rarely used Logical Symbols

These symbols are sorted by their Unicode value:

- U+00B7 · MIDDLE DOT, an outdated way for denoting AND, still in use in electronics; for example "A · B" is the same as "A & B"
- · : Center dot with a line above it; Outdated way for denoting NAND, for example "A·B" is the same as "A NAND B" or "A | B" or "¬ (A & B)". See also Unicode U+22C5 · DOT OPERATOR.
- U+0305 ̅ COMBINING OVERLINE, used as abbreviation for standard numerals (Typographical Number Theory). For example, using HTML style "4̅" is a shorthand for the standard numeral "SSSS0".
- Overline is also a rarely used format for denoting Gödel numbers: for example, "A V B" says the Gödel number of "(A V B)".
- Overline is also an outdated way for denoting negation, still in use in electronics: for example, "A V B" is the same as "¬(A V B)".
- U+2191 ↑ UPWARDS ARROW or U+007C | VERTICAL LINE: Sheffer stroke,

the sign for the NAND operator.

- U+2201 ∁ COMPLEMENT
- U+2204 ∄ THERE DOES NOT EXIST: strike out existential quantifier same as "¬∃"
- U+2234 ∴ THEREFORE: Therefore
- U+2235 ∵ BECAUSE: because
- U+22A7 ⊧ MODELS: is a model of
- U+22A8 ⊨ TRUE: is true of
- U+22AC ⊬ DOES NOT PROVE: negated ⊢, the sign for "does not prove", for example *T* ⊬ *P* says "*P* is not a theorem of *T*"
- U+22AD ⊭ NOT TRUE: is not true of
- U+22BC ⊼ NAND: NAND operator. In HTML, it can also be produced by <span style="text-decoration: overline">∧</span>: ∧
- U+22BD ⊽ NOR: NOR operator. In HTML, it can also be produced by <span style="text-decoration: overline">∨</span>: ∨
- U+25C7 ◇ WHITE DIAMOND: modal operator for "it is possible that", "it is not necessarily not" or rarely "it is not provable not" (in most modal logics it is defined as "¬¬")
- U+22C6 ⋆ STAR OPERATOR: usually used for ad-hoc operators
- U+22A5 ⊥ UP TACK or U+2193 ↓ DOWNWARDS ARROW: Webb-operator or Peirce arrow, the sign for NOR. Confusingly, "⊥" is also the sign for contradiction or absurdity.
- U+2310 ⌐ REVERSED NOT SIGN
- U+231C TOP LEFT CORNER and U+231D TOP RIGHT CORNER: corner quotes, also called "Quine quotes"; for quasi-quotation, i.e. quoting specific context of unspecified ("variable") expressions; also used for denoting Gödel number; for example "G" denotes the Gödel number of G. (Typographical note: although the quotes appears as a "pair" in unicode (231C and 231D), they are not symmetrical in some fonts. And in some fonts (for example Arial) they are only symmetrical in certain sizes. Alternatively the quotes can be rendered as ⌈ and ⌉ (U+2308 and U+2309) or by using a negation symbol and a reversed negation symbol ⌐ ¬ in superscript mode.)
- U+25FB WHITE MEDIUM SQUARE or U+25A1 □ WHITE SQUARE: modal operator for "it is necessary that" (in modal logic), or "it is provable that" (in provability logic), or "it is obligatory that" (in deontic logic), or "it is believed that" (in doxastic logic); also as empty clause (alternatives: ∅ and ⊥).

Note that the following operators are rarely supported by natively installed fonts. If you wish to use these in a web page, you should always embed the necessary fonts so the page viewer can see the web page without having the necessary fonts installed in their computer.

- U+27E1 ⟡ WHITE CONCAVE-SIDED DIAMOND

- U+27E2 WHITE CONCAVE-SIDED DIAMOND WITH LEFTWARDS TICK: modal operator for was never
- U+27E3 WHITE CONCAVE-SIDED DIAMOND WITH RIGHTWARDS TICK: modal operator for will never be
- U+27E4 WHITE SQUARE WITH LEFTWARDS TICK: modal operator for was always
- U+27E5 WHITE SQUARE WITH RIGHTWARDS TICK: modal operator for will always be
- U+297D ⥽ RIGHT FISH TAIL: sometimes used for "relation", also used for denoting various ad hoc relations (for example, for denoting "witnessing" in the context of Rosser's trick) The fish hook is also used as strict implication by C.I.Lewis $p ⥽ q \equiv \Box(p \rightarrow q)$, the corresponding LaTeX macro is \strictif. See here for an image of glyph. Added to Unicode 3.2.0.
- U+2A07 ⩇ TWO LOGICAL AND OPERATOR

REFERENCES

1. Walicki, Michał (2011), Introduction to Mathematical Logic, Singapore: World Scientific Publishing, ISBN 978-981-4343-87-9.
2. Boolos, George; Burgess, John; Jeffrey, Richard (2002), Computability and Logic (4th ed.), Cambridge: Cambridge University Press, ISBN 978-0-521-00758-0.
3. Crossley, J.N.; Ash, C.J.; Brickhill, C.J.; Stillwell, J.C.; Williams, N.H. (1972), What is mathematical logic?, London-Oxford-New York: Oxford University Press, ISBN 0-19-888087-1, Zbl 0251.02001.
4. Enderton, Herbert (2001), A mathematical introduction to logic (2nd ed.), Boston, MA: Academic Press, ISBN 978-0-12-238452-3.
5. Hamilton, A.G. (1988), Logic for Mathematicians (2nd ed.), Cambridge: Cambridge University Press, ISBN 978-0-521-36865-0.
6. Ebbinghaus, H.-D.; Flum, J.; Thomas, W. (1994), Mathematical Logic (2nd ed.), New York: Springer, ISBN 0-387-94258-0.
7. Katz, Robert (1964), Axiomatic Analysis, Boston, MA: D. C. Heath and Company.
8. Mendelson, Elliott (1997), Introduction to Mathematical Logic (4th ed.), London: Chapman & Hall, ISBN 978-0-412-80830-2.
9. Rautenberg, Wolfgang (2010), A Concise Introduction to Mathematical Logic (3rd ed.), New York: Springer Science+Business Media, doi:10.1007/978-1-4419-1221-3, ISBN 978-1-4419-1220-6.
10. Schwichtenberg, Helmut (2003–2004), Mathematical Logic (PDF), Munich, Germany: Mathematisches Institut der Universität München, retrieved 2016-02-24.
11. Shawn Hedman, A first course in logic: an introduction to model theory, proof theory, computability, and complexity, Oxford University Press, 2004, ISBN 0-19-852981-3. Covers logics in close relation with computability theory and complexity theory

12. van Dalen, Dirk (2013), Logic and Structure, Berlin: Springer-Verlag, ISBN 978-1-4471-4557-8.
13. Andrews, Peter B. (2002), An Introduction to Mathematical Logic and Type Theory: To Truth Through Proof (2nd ed.), Boston: Kluwer Academic Publishers, ISBN 978-1-4020-0763-7.
14. Barwise, Jon, ed. (1989). Handbook of Mathematical Logic. Studies in Logic and the Foundations of Mathematics. North Holland. ISBN 978-0-444-86388-1..
15. Hodges, Wilfrid (1997), A shorter model theory, Cambridge: Cambridge University Press, ISBN 978-0-521-58713-6.
16. Jech, Thomas (2003), Set Theory: Millennium Edition, Springer Monographs in Mathematics, Berlin, New York: Springer-Verlag, ISBN 978-3-540-44085-7.
17. Shoenfield, Joseph R. (2001) [1967], Mathematical Logic (2nd ed.), A K Peters, ISBN 978-1-56881-135-2.
18. Troelstra, Anne Sjerp; Schwichtenberg, Helmut (2000), Basic Proof Theory, Cambridge Tracts in Theoretical Computer Science (2nd ed.), Cambridge: Cambridge University Press, ISBN 978-0-521-77911-1.
19. Cohen, P. J. (1966), Set Theory and the Continuum Hypothesis, Menlo Park, CA: W. A. Benjamin.
20. Cohen, Paul Joseph (2008) [1966]. Set theory and the continuum hypothesis. Mineola, New York: Dover Publications. ISBN 978-0-486-46921-8..
21. J.D. Sneed, The Logical Structure of Mathematical Physics. Reidel, Dordrecht, 1971 (revised edition 1979).
22. Davis, Martin (1973), "Hilbert's tenth problem is unsolvable", The American Mathematical Monthly, The American Mathematical Monthly, Vol. 80, No. 3, 80 (3): 233–269, doi:10.2307/2318447, JSTOR 2318447, reprinted as an appendix in Martin Davis, Computability and Unsolvability, Dover reprint 1982. JStor
23. Felscher, Walter (2000), "Bolzano, Cauchy, Epsilon, Delta", The American Mathematical Monthly, The American Mathematical Monthly, Vol. 107, No. 9, 107 (9): 844–862, doi:10.2307/2695743, JSTOR 2695743. JSTOR(subscription required)
24. Ferreirós, José (2001), "The Road to Modern Logic-An Interpretation", Bulletin of Symbolic Logic, The Bulletin of Symbolic Logic, Vol. 7, No. 4, 7 (4): 441–484, doi:10.2307/2687794, JSTOR 2687794.
25. Hamkins, Joel David; Benedikt Löwe, "The modal logic of forcing", Transactions of the American Mathematical Society, arXiv:math/0509616, doi:10.1090/s0002-9947-07-04297-3
26. Katz, Victor J. (1998), A History of Mathematics, Addison–Wesley, ISBN 0-321-01618-1.
27. Morley, Michael (1965), "Categoricity in Power", Transactions of the American Mathematical Society, Transactions of the American Mathematical Society, Vol. 114, No. 2, 114 (2): 514–538, doi:10.2307/1994188, JSTOR 1994188.
28. Soare, Robert I. (1996), "Computability and recursion", Bulletin of Symbolic

Logic, The Bulletin of Symbolic Logic, Vol. 2, No. 3, 2 (3): 284–321, doi:10.2307/420992, JSTOR 420992.

29. Solovay, Robert M. (1976), "Provability Interpretations of Modal Logic", Israel Journal of Mathematics, 25 (3–4): 287–304, doi:10.1007/BF02757006.
30. Woodin, W. Hugh (2001), "The Continuum Hypothesis, Part I", Notices of the American Mathematical Society, 48 (6). PDF
31. Burali-Forti, Cesare (1897), A question on transfinite numbers, reprinted in van Heijenoort 1976, pp. 104–111.
32. Dedekind, Richard (1872), Stetigkeit und irrationale Zahlen. English translation of title: "Consistency and irrational numbers".
33. Dedekind, Richard (1888), Was sind und was sollen die Zahlen? Two English translations:
34. 1963 (1901). Essays on the Theory of Numbers. Beman, W. W., ed. and trans. Dover.
35. 1996. In From Kant to Hilbert: A Source Book in the Foundations of Mathematics, 2 vols, Ewald, William B., ed., Oxford University Press: 787–832.
36. Fraenkel, Abraham A. (1922), "Der Begriff 'definit' und die Unabhängigkeit des Auswahlsaxioms", Sitzungsberichte der Preussischen Akademie der Wissenschaften, Physikalisch-mathematische Klasse, pp. 253–257 (German), reprinted in English translation as «The notion of ‹definite› and the independence of the axiom of choice», van Heijenoort 1976, pp. 284–289.
37. Frege, Gottlob (1879), Begriffsschrift, eine der arithmetischen nachgebildete Formelsprache des reinen Denkens. Halle a. S.: Louis Nebert. Translation: Concept Script, a formal language of pure thought modelled upon that of arithmetic, by S. Bauer-Mengelberg in Jean Van Heijenoort, ed., 1967. From Frege to Gödel: A Source Book in Mathematical Logic, 1879–1931. Harvard University Press.
38. Frege, Gottlob (1884), Die Grundlagen der Arithmetik: eine logisch-mathematische Untersuchung über den Begriff der Zahl. Breslau: W. Koebner. Translation: J. L. Austin, 1974. The Foundations of Arithmetic: A logico-mathematical enquiry into the concept of number, 2nd ed. Blackwell.
39. Gentzen, Gerhard (1936), "Die Widerspruchsfreiheit der reinen Zahlentheorie", Mathematische Annalen, 112: 132–213, doi:10.1007/BF01565428, reprinted in English translation in Gentzen's Collected works, M. E. Szabo, ed., North-Holland, Amsterdam, 1969.
40. Gödel, Kurt (1929), Über die Vollständigkeit des Logikkalküls, doctoral dissertation, University Of Vienna. English translation of title: "Completeness of the logical calculus".
41. Gödel, Kurt (1930), "Die Vollständigkeit der Axiome des logischen Funktionenkalküls", Monatshefte für Mathematik und Physik, 37: 349–360, doi:10.1007/BF01696781. English translation of title: "The completeness of the axioms of the calculus of logical functions".
42. Gödel, Kurt (1931), "Über formal unentscheidbare Sätze der Principia Mathematica

und verwandter Systeme I", Monatshefte für Mathematik und Physik, 38 (1): 173–198, doi:10.1007/BF01700692, see On Formally Undecidable Propositions of Principia Mathematica and Related Systems for details on English translations.

43. Gödel, Kurt (1958), "Über eine bisher noch nicht benützte Erweiterung des finiten Standpunktes", Dialectica. International Journal of Philosophy, 12 (3–4): 280–287, doi:10.1111/j.1746-8361.1958.tb01464.x, reprinted in English translation in Gödel's Collected Works, vol II, Solomon Feferman et al., eds. Oxford University Press, 1990.
44. van Heijenoort, Jean, ed. (1976) [1967], From Frege to Gödel: A Source Book in Mathematical Logic, 1879–1931 (3rd ed.), Cambridge, Mass: Harvard University Press, ISBN 0-674-32449-8, (pbk.)
45. Hilbert, David (1899), Grundlagen der Geometrie, Leipzig: Teubner, English 1902 edition (The Foundations of Geometry) republished 1980, Open Court, Chicago.
46. David, Hilbert (1929), «Probleme der Grundlegung der Mathematik", Mathematische Annalen, 102: 1–9, doi:10.1007/BF01782335. Lecture given at the International Congress of Mathematicians, 3 September 1928. Published in English translation as "The Grounding of Elementary Number Theory", in Mancosu 1998, pp. 266–273.
47. Hilbert, David; Bernays, Paul (1934). Grundlagen der Mathematik. I. Die Grundlehren der mathematischen Wissenschaften. 40. Berlin, New York: Springer-Verlag. ISBN 978-3-540-04134-4. JFM 60.0017.02. MR 0237246.
48. Kleene, Stephen Cole (1943), "Recursive Predicates and Quantifiers", American Mathematical Society Transactions, Transactions of the American Mathematical Society, Vol. 53, No. 1, 54 (1): 41–73, doi:10.2307/1990131, JSTOR 1990131.
49. Lobachevsky, Nikolai (1840), Geometrishe Untersuchungen zur Theorie der Parellellinien (German). Reprinted in English translation as «Geometric Investigations on the Theory of Parallel Lines» in Non-Euclidean Geometry, Robert Bonola (ed.), Dover, 1955. ISBN 0-486-60027-0
50. Löwenheim, Leopold (1915), «Über Möglichkeiten im Relativkalkül", Mathematische Annalen, 76 (4): 447–470, doi:10.1007/BF01458217, ISSN 0025-5831 (German). Translated as "On possibilities in the calculus of relatives" in Jean van Heijenoort, 1967. A Source Book in Mathematical Logic, 1879–1931. Harvard Univ. Press: 228–251.
51. Mancosu, Paolo, ed. (1998), From Brouwer to Hilbert. The Debate on the Foundations of Mathematics in the 1920s, Oxford: Oxford University Press.
52. Pasch, Moritz (1882), Vorlesungen über neuere Geometrie.
53. Peano, Giuseppe (1889), Arithmetices principia, nova methodo exposita (Latin), excerpt reprinted in English translation as "The principles of arithmetic, presented by a new method", van Heijenoort 1976, pp. 83 97.
54. Richard, Jules (1905), "Les principes des mathématiques et le problème des ensembles", Revue générale des sciences pures et appliquées, 16: 541 (French), reprinted in English translation as "The principles of mathematics and the problems of sets", van Heijenoort 1976, pp. 142–144.

55. Skolem, Thoralf (1920), "Logisch-kombinatorische Untersuchungen über die Erfüllbarkeit oder Beweisbarkeit mathematischer Sätze nebst einem Theoreme über dichte Mengen", Videnskapsselskapet Skrifter, I. Matematisk-naturvidenskabelig Klasse, 6: 1–36.

56. Tarski, Alfred (1948), A decision method for elementary algebra and geometry, Santa Monica, California: RAND Corporation

57. Turing, Alan M. (1939), "Systems of Logic Based on Ordinals", Proceedings of the London Mathematical Society, 45 (2): 161–228, doi:10.1112/plms/s2-45.1.161

58. Zermelo, Ernst (1904), «Beweis, daß jede Menge wohlgeordnet werden kann", Mathematische Annalen, 59 (4): 514–516, doi:10.1007/BF01445300 (German), reprinted in English translation as "Proof that every set can be well-ordered", van Heijenoort 1976, pp. 139–141.

59. Zermelo, Ernst (1908a), «Neuer Beweis für die Möglichkeit einer Wohlordnung", Mathematische Annalen, 65: 107–128, doi:10.1007/BF01450054, ISSN 0025-5831 (German), reprinted in English translation as "A new proof of the possibility of a well-ordering", van Heijenoort 1976, pp. 183–198.

60. Zermelo, Ernst (1908b), «Untersuchungen über die Grundlagen der Mengenlehre", Mathematische Annalen, 65 (2): 261–281, doi:10.1007/BF01449999.

61. Goldrei, Derek (2005), Propositional and Predicate Calculus: A Model of Argument: A Model of Argument, Springer, p. 265, ISBN 9781846282294.

62. Szmielew, W. (1955), "Elementary properties of Abelian groups", Fundamenta Mathematicae, 41: 203–271, MR 0072131.

63. Ax, James; Kochen, Simon (1965), "Diophantine problems over local fields. II. A complete set of axioms for p-adic number theory.", Amer. J. Math., The Johns Hopkins University Press, 87 (3): 631–648, doi:10.2307/2373066, JSTOR 2373066, MR 018493

64. "Named character references". HTML 5.1 Nightly. W3C. Retrieved 9 September 2015.

65. Although this character is available in LaTeX, the MediaWiki TeX system doesn't support this character.

66. Brody, Baruch A. (1973), Logic: theoretical and applied, Prentice-Hall, p. 93, ISBN 9780135401460, We turn now to the second of our connective symbols, the centered dot, which is called the conjunction sign.

67. Quine, W.V. (1981): Mathematical Logic, §6

68. Hintikka, Jaakko (1998), The Principles of Mathematics Revisited, Cambridge University Press, p. 113, ISBN 9780521624985.

69. Hermes, Hans. Einführung in die mathematische Logik: klassische Prädikatenlogik. Springer-Verlag, 2013.ss

CHAPTER 2

Set Theory

Chapter Contents

- **History**
- **Basic Concepts and Notation**
- **Category Theory**
- **Relational Model**
- **Relational Algebra**
- **Relational Calculus**

Set theory is the mathematical theory of well-determined collections, called sets, of objects that are called members, or elements, of the set. Pure set theory deals exclusively with sets, so the only sets under consideration are those whose members are also sets.

Set theory is a branch of mathematical logic that studies sets, which informally are collections of objects. Although any type of object can be collected into a set, set theory is applied most often to objects that are relevant to mathematics. The language of set theory can be used in the definitions of nearly all mathematical objects.

The modern study of set theory was initiated by Georg Cantor and Richard Dedekind in the 1870s. After the discovery of paradoxes in naive set theory, such as the Russell›s paradox, numerous axiom systems were proposed in the early twentieth century, of which the Zermelo–Fraenkel axioms, with the axiom of choice, are the best-known.

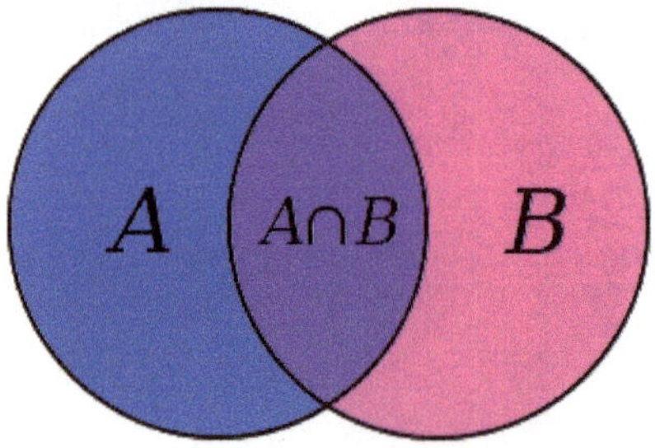

A Venn diagram illustrating the intersection of two sets.

Set theory is commonly employed as a foundational system for mathematics, particularly in the form of Zermelo–Fraenkel set theory with the axiom of choice. Beyond its foundational role, set theory is a branch of mathematics in its own right, with an active research community. Contemporary research into set theory includes a diverse collection of topics, ranging from the structure of the real number line to the study of the consistency of large cardinals.

HISTORY

Georg Cantor.

Mathematical topics typically emerge and evolve through interactions among many researchers. Set theory, however, was founded by a single paper in 1874 by Georg Cantor: "On a Property of the Collection of All Real Algebraic Numbers".

Since the 5th century BC, beginning with Greek mathematician Zeno of Elea in the West and early Indian mathematicians in the East, mathematicians had struggled with the concept of infinity. Especially notable is the work of Bernard Bolzano in the first half of the 19th century. Modern understanding of infinity began in 1867–71, with Cantor's work on number theory. An 1872 meeting between Cantor and Richard Dedekind influenced Cantor's thinking and culminated in Cantor's 1874 paper.

Cantor›s work initially polarized the mathematicians of his day. While Karl Weierstrass and Dedekind supported Cantor, Leopold Kronecker, now seen as a founder of mathematical constructivism, did not. Cantorian set theory eventually became widespread, due to the utility of Cantorian concepts, such as one-to-one correspondence among sets, his proof that there are more real numbers than integers, and the "infinity of infinities" ("Cantor›s paradise) resulting from the power set operation. This utility of set theory led to the article "Mengenlehre contributed in 1898 by Arthur Schoenflies to Klein›s encyclopedia.

The next wave of excitement in set theory came around 1900, when it was discovered that some interpretations of Cantorian set theory gave rise to several contradictions, called antinomies or paradoxes. Bertrand Russell and Ernst Zermelo independently found the simplest and best known paradox, now called Russell›s paradox: consider the set of all sets that are not members of themselves, which leads to a contradiction since it must be a member of itself, and not a member of itself. In 1899 Cantor had himself posed the question What is the cardinal number of the set of all sets?", and obtained a related paradox. Russell used his paradox as a theme in his 1903 review of continental mathematics in his The Principles of Mathematics.

In 1906 English readers gained the book Theory of Sets of Points by William Henry Young and his wife Grace Chisholm Young, published by Cambridge University Press.

The momentum of set theory was such that debate on the paradoxes did not lead to its abandonment. The work of Zermelo in 1908 and Abraham Fraenkel in 1922 resulted in the set of axioms ZFC, which became the most commonly used set of axioms for set theory. The work of analysts such as Henri Lebesgue demonstrated the great mathematical utility of set theory, which has since become woven into the fabric of modern mathematics. Set theory is commonly used as a foundational system, although in some areas category theory is thought to be a preferred foundation.

BASIC CONCEPTS AND NOTATION

Set theory begins with a fundamental binary relation between an object o and a set A. If o is a **member** (or **element**) of A, the notation $o \in A$ is used. Since sets are objects, the membership relation can relate sets as well.

A derived binary relation between two sets is the subset relation, also called **set inclusion**. If all the members of set A are also members of set B, then A is a **subset** of B, denoted $A \subseteq B$. For example, $\{1, 2\}$ is a subset of $\{1, 2, 3\}$, and so is $\{2\}$ but $\{1, 4\}$ is not. As insinuated from this definition, a set is a subset of itself. For cases where this possibility is unsuitable or would make sense to be rejected, the term **proper subset** is defined. A is called a **proper subset** of B if and only if A is a subset of B, but A is not equal to B. Note also that 1 and 2 and 3 are members (elements) of set {1, 2, 3} , but are not subsets, and the subsets are, in turn, not as such members of the set.

Just as arithmetic features binary operations on numbers, set theory features binary operations on sets. The:

- **Union** of the sets A and B, denoted A ∪ B, is the set of all objects that are a member of A, or B, or both. The union of {1, 2, 3} and {2, 3, 4} is the set {1, 2, 3, 4} .
- **Intersection** of the sets A and B, denoted A ∩ B, is the set of all objects that are members of both A and B. The intersection of {1, 2, 3} and {2, 3, 4} is the set {2, 3} .
- **Set difference** of U and A, denoted U \ A, is the set of all members of U that are not members of A. The set difference {1, 2, 3} \ {2, 3, 4} is {1} , while, conversely, the set difference {2, 3, 4} \ {1, 2, 3} is {4} . When A is a subset of U, the set difference U \ A is also called the **complement** of A in U. In this case, if the choice of U is clear from the context, the notation A^c is sometimes used instead of U \ A, particularly if U is a universal set as in the study of Venn diagrams.
- **Symmetric difference** of sets A and B, denoted A △ B or A ⊖ B, is the set of all objects that are a member of exactly one of A and B(elements which are in one of the sets, but not in both). For instance, for the sets {1, 2, 3} and {2, 3, 4} , the symmetric difference set is {1, 4}. It is the set difference of the union and the intersection, (A ∪ B) \ (A ∩ B) or (A \ B) ∪ (B \ A).
- **Cartesian product** of A and B, denoted A × B, is the set whose members are all possible ordered pairs (a, b) where a is a member of A and bis a member of B. The cartesian product of {1, 2} and {red, white} is {(1, red), (1, white), (2, red), (2, white)}.
- **Power set** of a set A is the set whose members are all possible subsets of A. For example, the power set of {1, 2} is { {}, {1}, {2}, {1, 2} } .

Some basic sets of central importance are the empty set (the unique set containing no elements; occasionally called the null set though this name is ambiguous), the set of natural numbers, and the set of real numbers.

Some Ontology

A set is pure if all of its members are sets, all members of its members are sets, and so on. For example, the set {{}} containing only the empty set is a nonempty pure set. In modern set theory, it is common to restrict attention to the **von Neumann universe** of pure sets, and many systems of axiomatic set theory are designed to axiomatize the pure sets only. There are many technical advantages to this restriction, and little generality is lost, because essentially all mathematical concepts can be modeled by pure sets. Sets in the von Neumann universe are organized into a cumulative hierarchy, based on how deeply their members, members of members, etc. are nested. Each set in this hierarchy is assigned (by transfinite recursion) an ordinal number α, known as its **rank**. The rank of a pure set X is defined to be the least upper bound of all successors of ranks of members of X. For example, the empty set is assigned rank 0, while the set {{}} containing only the empty set is assigned rank 1. For each ordinal α, the set V_α is defined to consist of all pure sets with rank less than α. The entire von Neumann universe is denoted V.

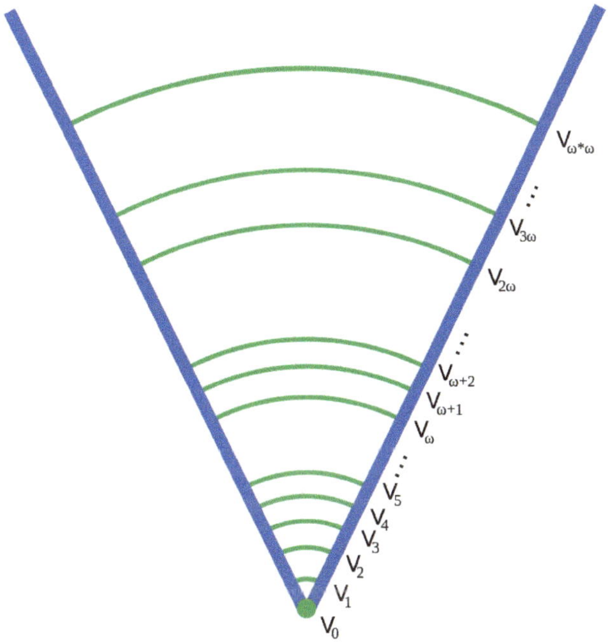

An initial segment of the von Neumann hierarchy.

Axiomatic Set Theory

Elementary set theory can be studied informally and intuitively, and so can be taught in primary schools using Venn diagrams. The intuitive approach tacitly assumes that a set may be formed from the class of all objects satisfying any particular defining condition. This assumption gives rise to paradoxes, the simplest and best known of which are Russell's paradox and the Burali-Forti paradox. Axiomatic set theory was originally devised to rid set theory of such paradoxes.

The most widely studied systems of axiomatic set theory imply that all sets form a cumulative hierarchy. Such systems come in two flavors, those whose ontology consists of:

- Sets alone. This includes the most common axiomatic set theory, **Zermelo–Fraenkel set theory (ZFC)**, which includes the axiom of choice. Fragments of ZFC include:
  - Zermelo set theory, which replaces the axiom schema of replacement with that of separation;

- General set theory, a small fragment of Zermelo set theory sufficient for the Peano axioms and finite sets;
- Kripke–Platek set theory, which omits the axioms of infinity, powerset, and choice, and weakens the axiom schemata of separation and replacement.

- Sets and proper classes. These include Von Neumann–Bernays–Gödel set theory, which has the same strength as ZFC for theorems about sets alone, and Morse-Kelley set theory and Tarski–Grothendieck set theory, both of which are stronger than ZFC.

The above systems can be modified to allow **urelements**, objects that can be members of sets but that are not themselves sets and do not have any members.

The systems of **New Foundations NFU** (allowing urelements) and **NF** (lacking them) are not based on a cumulative hierarchy. NF and NFU include a "set of everything, relative to which every set has a complement. In these systems urelements matter, because NF, but not NFU, produces sets for which the axiom of choice does not hold.

Systems of constructive set theory, such as CST, CZF, and IZF, embed their set axioms in intuitionistic instead of classical logic. Yet other systems accept classical logic but feature a nonstandard membership relation. These include rough set theory and fuzzy set theory, in which the value of an atomic formula embodying the membership relation is not simply **True** or **False**. The Boolean-valued models of ZFC are a related subject.

An enrichment of ZFC called Internal Set Theory was proposed by Edward Nelson in 1977.

Applications

Many mathematical concepts can be defined precisely using only set theoretic concepts. For example, mathematical structures as diverse as graphs, manifolds, rings, and vector spaces can all be defined as sets satisfying various (axiomatic) properties. Equivalence and order relations are ubiquitous in mathematics, and the theory of mathematical relations can be described in set theory.

Set theory is also a promising foundational system for much of mathematics. Since the publication of the first volume of Principia Mathematica, it has been claimed that most or even all mathematical theorems can be derived using an aptly designed set of axioms for set theory, augmented with many definitions, using first or second order logic. For example, properties of the natural and real numbers can be derived within set theory, as each number system can be identified with a set of equivalence classes under a suitable equivalence relation whose field is some infinite set.

Set theory as a foundation for mathematical analysis, topology, abstract algebra, and discrete mathematics is likewise uncontroversial; mathematicians accept that (in principle) theorems in these areas can be derived from the relevant definitions and the axioms of set theory. Few full derivations of complex mathematical theorems from set theory have been formally verified, however, because such formal derivations are often much longer than the natural language proofs mathematicians commonly present. One verification project, Metamath, includes human-written, computer-verified derivations of more than 12,000 theorems starting from ZFC set theory, first order logic and propositional logic.

Areas of Study

Set theory is a major area of research in mathematics, with many interrelated subfields.

Combinatorial Set Theory

Infinitary combinatorics

Combinatorial set theory concerns extensions of finite combinatorics to infinite sets. This includes the study of cardinal arithmetic and the study of extensions of Ramsey›s theorem such as the Erdős–Rado theorem.

Descriptive Set Theory

Descriptive set theory is the study of subsets of the real line and, more generally, subsets of Polish spaces. It begins with the study of pointclasses in the Borel hierarchy and extends to the study of more complex hierarchies such as the projective hierarchy and the Wadge hierarchy. Many properties of Borel sets can be established in ZFC, but proving these properties hold for more complicated sets requires additional axioms related to determinacy and large cardinals.

The field of effective descriptive set theory is between set theory and recursion theory. It includes the study of lightface pointclasses, and is closely related to hyperarithmetical theory. In many cases, results of classical descriptive set theory have effective versions; in some cases, new results are obtained by proving the effective version first and then extending (relativizing) it to make it more broadly applicable.

A recent area of research concerns Borel equivalence relations and more complicated definable equivalence relations. This has important applications to the study of invariants in many fields of mathematics.

Fuzzy Set Theory

In set theory as Cantor defined and Zermelo and Fraenkel axiomatized, an object is either a member of a set or not. In fuzzy set theory this condition was relaxed by Lotfi A. Zadeh so an object has a degree of membership in a set, a number between 0 and 1. For example, the degree of membership of a person in the set of "tall people" is more flexible than a simple yes or no answer and can be a real number such as 0.75.

Inner Model Theory

An **inner model** of Zermelo–Fraenkel set theory (ZF) is a transitive class that includes all the ordinals and satisfies all the axioms of ZF. The canonical example is the constructible universe L developed by Gödel. One reason that the study of inner models is of interest is that it can be used to prove consistency results. For example, it can be shown that regardless of whether a model V of ZF satisfies the continuum hypothesis or the axiom of choice, the inner model L constructed inside the original model will satisfy both the generalized continuum hypothesis and the axiom of choice. Thus the assumption that ZF is consistent (has at least one model) implies that ZF together with these two principles is consistent.

The study of inner models is common in the study of determinacy and large cardinals, especially when considering axioms such as the axiom of determinacy that contradict the axiom of choice. Even if a fixed model of set theory satisfies the axiom of choice, it is possible for an inner model to fail to satisfy the axiom of choice. For example, the existence of sufficiently large cardinals implies that there is an inner model satisfying the axiom of determinacy (and thus not satisfying the axiom of choice).

Large Cardinals

A **large cardinal** is a cardinal number with an extra property. Many such properties are studied, including inaccessible cardinals, measurable cardinals, and many more. These properties typically imply the cardinal number must be very large, with the existence of a cardinal with the specified property unprovable in Zermelo-Fraenkel set theory.

Determinacy

Determinacy refers to the fact that, under appropriate assumptions, certain two-player games of perfect information are determined from the start in the sense that one player must have a winning strategy. The existence of these strategies has important consequences in descriptive set theory, as the assumption that a broader class of games is determined often implies that a broader class of sets will have a topological property. The axiom of determinacy (AD) is an important object of study; although incompatible with the axiom of choice, AD implies that all subsets of the real line are well behaved (in particular, measurable and with the perfect set property). AD can be used to prove that the Wadge degrees have an elegant structure.

Forcing

Paul Cohen invented the method of forcing while searching for a model of ZFC in which the continuum hypothesis fails, or a model of ZF in which the axiom of choice fails. Forcing adjoins to some given model of set theory additional sets in order to create a larger model with properties determined (i.e. "forced") by the construction and the original model. For example, Cohen's construction adjoins additional subsets of the natural numbers without changing any of the cardinal numbers of the original model. Forcing is also one of two methods for proving relative consistency by finitistic methods, the other method being Boolean-valued models.

Cardinal Invariants

A **cardinal invariant** is a property of the real line measured by a cardinal number. For example, a well-studied invariant is the smallest cardinality of a collection of meagre sets of reals whose union is the entire real line. These are invariants in the sense that any two isomorphic models of set theory must give the same cardinal for each invariant.

Many cardinal invariants have been studied, and the relationships between them are often complex and related to axioms of set theory.

Set-Theoretic Topology

Set-theoretic topology studies questions of general topology that are set-theoretic in nature or that require advanced methods of set theory for their solution. Many of these theorems are independent of ZFC, requiring stronger axioms for their proof. A famous problem is the normal Moore space question, a question in general topology that was the subject of intense research. The answer to the normal Moore space question was eventually proved to be independent of ZFC.

Objections to Set Theory as a Foundation for Mathematics

From set theory's inception, some mathematicians have objected to it as a foundation for mathematics. The most common objection to set theory, one Kronecker voiced in set theory's earliest years, starts from the constructivist view that mathematics is loosely related to computation. If this view is granted, then the treatment of infinite sets, both in naive and in axiomatic set theory, introduces into mathematics methods and objects that are not computable even in principle. The feasibility of constructivism as a substitute foundation for mathematics was greatly increased by Errett Bishop›s influential book Foundations of Constructive Analysis.

A different objection put forth by Henri Poincaré is that defining sets using the axiom schemas of specification and replacement, as well as the axiom of power set, introduces impredicativity, a type of circularity, into the definitions of mathematical objects. The scope of predicatively founded mathematics, while less than that of the commonly accepted Zermelo-Fraenkel theory, is much greater than that of constructive mathematics, to the point that Solomon Feferman has said that "all of scientifically applicable analysis can be developed [using predicative methods]".

Ludwig Wittgenstein condemned set theory. He wrote that "set theory is wrong", since it builds on the "nonsense" of fictitious symbolism, has "pernicious idioms", and that it is nonsensical to talk about "all numbers". Wittgenstein's views about the foundations of mathematics were later criticised by Georg Kreisel and Paul Bernays, and investigated by Crispin Wright, among others.

Category theorists have proposed topos theory as an alternative to traditional axiomatic set theory. Topos theory can interpret various alternatives to that theory, such as constructivism, finite set theory, and computable set theory. Topoi also give a natural setting for forcing and discussions of the independence of choice from ZF, as well as providing the framework for pointless topology and Stone spaces.

An active area of research is the univalent foundations arising from homotopy type theory. Here, sets may be defined as certain kinds of types, with universal properties of sets arising from higher inductive types. Principles such as the axiom of choice and the law of the excluded middle appear in a spectrum of different forms, some of which can be proven, others which correspond to the classical notions; this allows for a detailed discussion of the effect of these axioms on mathematics.

CATEGORY THEORY

Category theory formalizes mathematical structure and its concepts in terms of a collection of objects and of arrows (also called morphisms). A category has two basic properties: the ability to compose the arrows associatively and the existence of an identity arrow for each object. The language of category theory has been used to formalize concepts of other high-level abstractions such as sets, rings, and groups.

Several terms used in category theory, including the term morphism, are used differently from their uses in the rest of mathematics. In category theory, morphisms obey conditions specific to category theory itself.

Samuel Eilenberg and Saunders Mac Lane introduced the concepts of categories, functors, and natural transformations in 1942–45 in their study of algebraic topology, with the goal of understanding the processes that preserve mathematical structure.

Category theory has practical applications in programming language theory, in particular for the study of monads in functional programming.

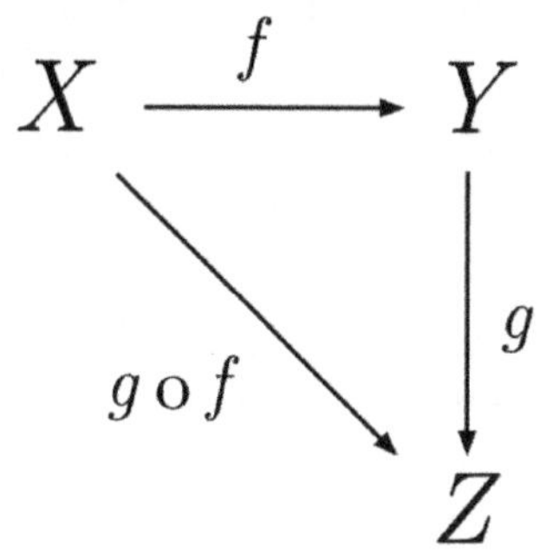

Schematic representation of a category with objects X, Y, Z and morphisms f, g, g ∘ f. (The category's three identity morphisms 1_X, 1_Y and 1_Z, if explicitly represented, would appear as three arrows, next to the letters X, Y, and Z, respectively, each having as its "shaft" a circular arc measuring almost 360 degrees.)

Categories represent abstraction of other mathematical concepts. Many areas of mathematics can be formalised by category theory as categories. Hence category theory uses abstraction to make it possible to state and prove many intricate and subtle mathematical results in these fields in a much simpler way.

A basic example of a category is the category of sets, where the objects are sets and the arrows are functions from one set to another. However, the objects of a category need not be sets, and the arrows need not be functions. Any way of formalising a mathematical concept such that it meets the basic conditions on the behaviour of objects and arrows is a valid category—and all the results of category theory apply to it.

The "arrows" of category theory are often said to represent a process connecting two objects, or in many cases a "structure-preserving" transformation connecting two objects. There are, however, many applications where much more abstract concepts are represented by objects and morphisms. The most important property of the arrows is that they can be "composed", in other words, arranged in a sequence to form a new arrow.

Applications of Categories

Categories now appear in many branches of mathematics, some areas of theoretical computer science where they can correspond to types or to database schemas, and mathematical physics where they can be used to describe vector spaces. Linear algebra can also be expressed in terms of categories of matrices.

Utility

Categories, objects, and morphisms

The study of categories is an attempt to axiomatically capture what is commonly found in various classes of related mathematical structures by relating them to the structure-preserving functions between them. A systematic study of category theory then allows us to prove general results about any of these types of mathematical structures from the axioms of a category.

Consider the following example. The class **Grp** of groups consists of all objects having a "group structure". One can proceed to prove theorems about groups by making logical deductions from the set of axioms. For example, it is immediately proven from the axioms that the identity element of a group is unique.

Instead of focusing merely on the individual objects (e.g., groups) possessing a given structure, category theory emphasizes the morphisms – the structure-preserving mappings – between these objects; by studying these morphisms, we are able to learn more about the structure of the objects. In the case of groups, the morphisms are the group homomorphisms. A group homomorphism between two groups preserves the group structure in a precise sense – it is a process taking one group to another, in a way that carries along information about the structure of the first group into the second group. The study of group homomorphisms then provides a tool for studying general properties of groups and consequences of the group axioms.

A similar type of investigation occurs in many mathematical theories, such as the study of continuous maps (morphisms) between topological spaces in topology (the associated category is called **Top**), and the study of smooth functions (morphisms) in manifold theory.

Not all categories arise as structure preserving (set) functions, however; the standard example is the category of homotopies between pointed topological spaces.

If one axiomatizes relations instead of functions, one obtains the theory of allegories.

Functors

A category is itself a type of mathematical structure, so we can look for "processes" which preserve this structure in some sense; such a process is called a functor.

Diagram chasing is a visual method of arguing with abstract "arrows" joined in diagrams. Functors are represented by arrows between categories, subject to specific defining commutativity conditions. Functors can define (construct) categorical diagrams and sequences (viz. Mitchell, 1965). A functor associates to every object of one category an object of another category, and to every morphism in the first category a morphism in the second.

In fact, what we have done is define a category of categories and functors – the objects are categories, and the morphisms (between categories) are functors.

By studying categories and functors, we are not just studying a class of mathematical structures and the morphisms between them; we are studying the relationships between various classes of mathematical structures. This is a fundamental idea, which first surfaced in algebraic topology. Difficult topological questions can be translated into algebraic questions which are often easier to solve. Basic constructions, such as the fundamental group or the fundamental groupoid of a topological space, can be expressed as functors to the category of groupoids in this way, and the concept is pervasive in algebra and its applications.

Natural Transformations

Abstracting yet again, some diagrammatic and/or sequential constructions are often "naturally related" – a vague notion, at first sight. This leads to the clarifying concept of natural transformation, a way to "map" one functor to another. Many important constructions in mathematics can be studied in this context. "Naturality" is a principle, like general covariance in physics, that cuts deeper than is initially apparent. An arrow between two functors is a natural transformation when it is subject to certain naturality or commutativity conditions.

Functors and natural transformations (‹naturality›) are the key concepts in category theory.

Categories, Objects, and Morphisms

Categories

A category C consists of the following three mathematical entities:

- A class ob(C), whose elements are called objects;
- A class hom(C), whose elements are called morphisms or maps or arrows. Each morphism ***f*** has a source object ***a*** and target object ***b***.
- The expression f : a → b, would be verbally stated as "f is a morphism from a to b".

- The expression **hom(*a*, *b*)** — alternatively expressed as **hom$_C$(*a*, *b*)**, **mor(*a*, *b*)**, or ***C*(*a*, *b*)** — denotes the hom-class of all morphisms from a to b.
- A binary operation ∘, called composition of morphisms, such that for any three objects a, b, and c, we have hom(b, c) × hom(a, b) → hom(a, c). The composition of f : a → b and g : b → c is written as g ∘ f or gf, governed by two axioms:
- Associativity: If f : a → b, g : b → c and h : c → d then h ∘ (g ∘ f) = (h ∘ g) ∘ f, and
- Identity: For every object x, there exists a morphism 1_x : x → x called the identity morphism for x, such that for every morphism f : a → b, we have 1_b ∘ f = f = f ∘ 1_a.

From the axioms, it can be proved that there is exactly one identity morphism for every object. Some authors deviate from the definition just given by identifying each object with its identity morphism.

Morphisms

Relations among morphisms (such as fg = h) are often depicted using commutative diagrams, with "points" (corners) representing objects and "arrows" representing morphisms.

Morphisms can have any of the following properties. A morphism f : a → b is a:

- monomorphism (or monic) if f ∘ g_1 = f ∘ g_2 implies g_1 = g_2 for all morphisms g_1, g_2 : x → a.
- epimorphism (or epic) if g_1 ∘ f = g_2 ∘ f implies g_1 = g_2 for all morphisms g_1, g_2 : b → x.
- bimorphism if f is both epic and monic.
- isomorphism if there exists a morphism g : b → a such that f ∘ g = 1_b and g ∘ f = 1_a.
- endomorphism if a = b. end(a) denotes the class of endomorphisms of a.
- automorphism if f is both an endomorphism and an isomorphism. aut(a) denotes the class of automorphisms of a.
- retraction if a right inverse of f exists, i.e. if there exists a morphism g : b → a with f ∘ g = 1_b.
- section if a left inverse of f exists, i.e. if there exists a morphism g : b → a with g ∘ f = 1_a.

Every retraction is an epimorphism, and every section is a monomorphism. Furthermore, the following three statements are equivalent:

- f is a monomorphism and a retraction;
- f is an epimorphism and a section;
- f is an isomorphism.

Functors

Functors are structure-preserving maps between categories. They can be thought of as morphisms in the category of all (small) categories.

A (**covariant**) functor F from a category C to a category D, written F : C → D, consists of:

- for each object x in C, an object F(x) in D; and
- for each morphism f : x → y in C, a morphism F(f) : F(x) → F(y),

such that the following two properties hold:

- For every object x in C, $F(1_x) = 1_{F(x)}$;
- For all morphisms f : x → y and g : y → z, F(g ∘ f) = F(g) ∘ F(f).

A **contravariant** functor F: C → D, is like a covariant functor, except that it "turns morphisms around" ("reverses all the arrows"). More specifically, every morphism f : x → y in C must be assigned to a morphism F(f) : F(y) → F(x) in D. In other words, a contravariant functor acts as a covariant functor from the opposite category C^{op} to D.

Natural Transformations

A natural transformation is a relation between two functors. Functors often describe natural constructions and natural transformations then describe natural homomorphisms between two such constructions. Sometimes two quite different constructions yield the same result; this is expressed by a natural isomorphism between the two functors.

If F and G are (covariant) functors between the categories C and D, then a natural transformation η from F to G associates to every object X in C a morphism η_X : F(X) → G(X) in D such that for every morphism f : X → Y in C, we have $\eta_Y \circ F(f) = G(f) \circ \eta_X$; this means that the following diagram is commutative:

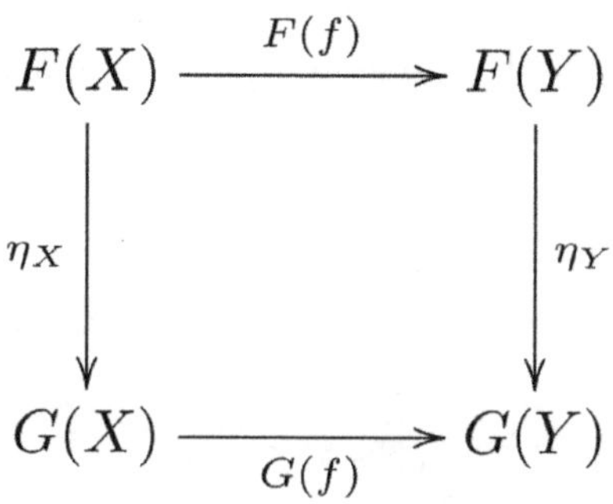

The two functors F and G are called naturally isomorphic if there exists a natural transformation from F to G such that η_X is an isomorphism for every object X in C.

Other Concepts

Universal Constructions, Limits, and Colimits

Using the language of category theory, many areas of mathematical study can be categorized. Categories include sets, groups and topologies.

Each category is distinguished by properties that all its objects have in common, such as the empty set or the product of two topologies, yet in the definition of a category, objects are considered atomic, i.e., we do not know whether an object A is a set, a topology, or any other abstract concept. Hence, the challenge is to define special objects without referring to the internal structure of those objects. To define the empty set without referring to elements, or the product topology without referring to open sets, one can characterize these objects in terms of their relations to other objects, as given by the morphisms of the respective categories. Thus, the task is to find universal properties that uniquely determine the objects of interest.

Indeed, it turns out that numerous important constructions can be described in a purely categorical way. The central concept which is needed for this purpose is called categorical limit, and can be dualized to yield the notion of a colimit.

Equivalent Categories

It is a natural question to ask: under which conditions can two categories be considered essentially the same, in the sense that theorems about one category can readily be transformed into theorems about the other category? The major tool one employs to describe such a situation is called equivalence of categories, which is given by appropriate functors between two categories. Categorical equivalence has found numerous applications in mathematics.

Further Concepts and Results

The definitions of categories and functors provide only the very basics of categorical algebra; additional important topics are listed below. Although there are strong interrelations between all of these topics, the given order can be considered as a guideline for further reading.

- The functor category D^C has as objects the functors from C to D and as morphisms the natural transformations of such functors. The Yoneda lemma is one of the most famous basic results of category theory; it describes representable functors in functor categories.
- Duality: Every statement, theorem, or definition in category theory has a dual which is essentially obtained by "reversing all the arrows". If one statement is true in a category C then its dual is true in the dual category C^{op}. This duality, which is transparent at the level of category theory, is often obscured in applications and can lead to surprising relationships.
- Adjoint functors: A functor can be left (or right) adjoint to another functor that maps in the opposite direction. Such a pair of adjoint functors typically arises

from a construction defined by a universal property; this can be seen as a more abstract and powerful view on universal properties.

Higher-Dimensional Categories

Many of the above concepts, especially equivalence of categories, adjoint functor pairs, and functor categories, can be situated into the context of higher-dimensional categories. Briefly, if we consider a morphism between two objects as a "process taking us from one object to another", then higher-dimensional categories allow us to profitably generalize this by considering "higher-dimensional processes".

For example, a (strict) 2-category is a category together with "morphisms between morphisms, i.e., processes which allow us to transform one morphism into another. We can then compose these bimorphisms both horizontally and vertically, and we require a 2-dimensional exchange law to hold, relating the two composition laws. In this context, the standard example is **Cat**, the 2-category of all (small) categories, and in this example, bimorphisms of morphisms are simply natural transformations of morphisms in the usual sense. Another basic example is to consider a 2-category with a single object; these are essentially monoidal categories. Bicategories are a weaker notion of 2-dimensional categories in which the composition of morphisms is not strictly associative, but only associative up to an isomorphism.

This process can be extended for all natural numbers n, and these are called n-categories. There is even a notion of ω-category corresponding to the ordinal number ω.

Higher-dimensional categories are part of the broader mathematical field of higher-dimensional algebra, a concept introduced by Ronald Brown. For a conversational introduction to these ideas, see John Baez, ‹A Tale of n-categories' (1996).

Historical Notes

In 1942–45, Samuel Eilenberg and Saunders Mac Lane introduced categories, functors, and natural transformations as part of their work in topology, especially algebraic topology. Their work was an important part of the transition from intuitive and geometric homology to axiomatichomology theory. Eilenberg and Mac Lane later wrote that their goal was to understand natural transformations. That required defining functors, which required categories.

Stanislaw Ulam, and some writing on his behalf, have claimed that related ideas were current in the late 1930s in Poland. Eilenberg was Polish, and studied mathematics in Poland in the 1930s. Category theory is also, in some sense, a continuation of the work of Emmy Noether (one of Mac Lane's teachers) in formalizing abstract processes; Noether realized that understanding a type of mathematical structure requires understanding the processes that preserve that structure. To achieve this understanding, Eilenberg and Mac Lane proposed an axiomatic formalization of the relation between structures and the processes that preserve them.

The subsequent development of category theory was powered first by the computational needs of homological algebra, and later by the axiomatic needs of algebraic geometry, the field most resistant to being grounded in either axiomatic set theory or the Russell-Whitehead view of united foundations. General category theory, an extension of universal algebra having many new features allowing for semantic flexibility and higher-order logic, came later; it is now applied throughout mathematics.

Certain categories called topoi (singular topos) can even serve as an alternative to axiomatic set theory as a foundation of mathematics. A topos can also be considered as a specific type of category with two additional topos axioms. These foundational applications of category theory have been worked out in fair detail as a basis for, and justification of, constructive mathematics. Topos theory is a form of abstract sheaf theory, with geometric origins, and leads to ideas such as pointless topology.

Categorical logic is now a well-defined field based on type theory for intuitionistic logics, with applications in functional programming and domain theory, where a cartesian closed category is taken as a non-syntactic description of a lambda calculus. At the very least, category theoretic language clarifies what exactly these related areas have in common (in some abstract sense).

Category theory has been applied in other fields as well. For example, John Baez has shown a link between Feynman diagrams in Physics and monoidal categories. Another application of category theory, more specifically: topos theory, has been made in mathematical music theory, see for example the book The Topos of Music, Geometric Logic of Concepts, Theory, and Performance by Guerino Mazzola.

More recent efforts to introduce undergraduates to categories as a foundation for mathematics include those of William Lawvere and Rosebrugh (2003) and Lawvere and Stephen Schanuel (1997) and Mirroslav Yotov (2012).

RELATIONAL MODEL

The **relational model (RM)** for database management is an approach to managing data using a structure and language consistent with first-order predicate logic, first described in 1969 by Edgar F. Codd, where all data is represented in terms of tuples, grouped into relations. A database organized in terms of the relational model is a relational database.

The purpose of the relational model is to provide a declarative method for specifying data and queries: users directly state what information the database contains and what information they want from it, and let the database management system software take care of describing data structures for storing the data and retrieval procedures for answering queries.

Most relational databases use the SQL data definition and query language; these systems implement what can be regarded as an engineering approximation to the relational model. A table in an SQL database schema corresponds to a predicate variable; the contents of a table to a relation; key constraints, other constraints, and SQL queries correspond to predicates. However, SQL databases deviate from the relational model in many details, and Codd fiercely argued against deviations that compromise the original principles.

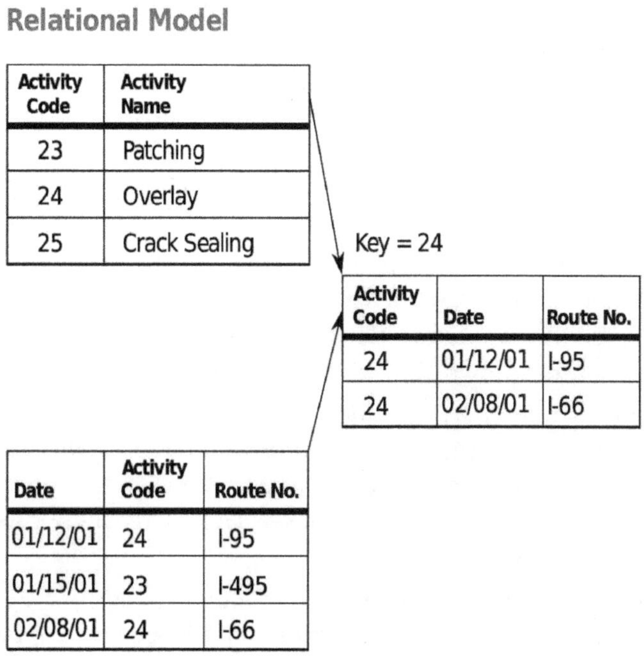

Diagram of an example database according to the relational model

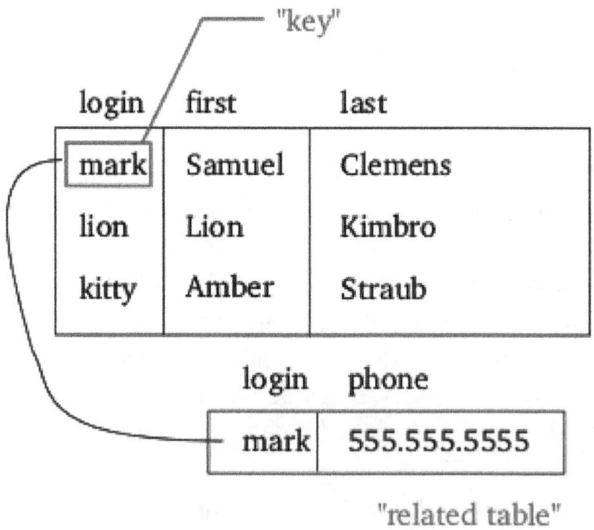

In the relational model, related records are linked together with a "key".

The relational model›s central idea is to describe a database as a collection of predicates over a finite set of predicate variables, describing constraints on the possible values and combinations of values. The content of the database at any given time is a finite (logical) model of the database, i.e. a set of relations, one per predicate variable, such that all predicates are satisfied. A request for information from the database (a database query) is also a predicate.

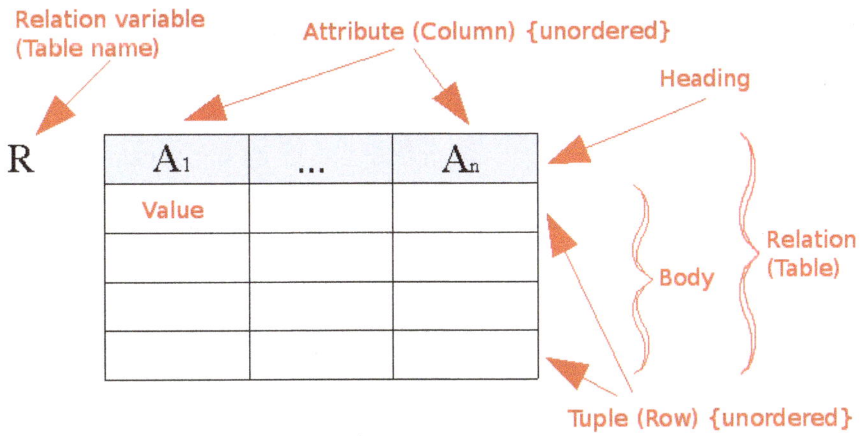

Relational model concepts.

Alternatives

Other models are the hierarchical model and network model. Some systems using these older architectures are still in use today in data centers with high data volume needs, or where existing systems are so complex and abstract it would be cost-prohibitive to migrate to systems employing the relational model. Also of note are newer object-oriented databases.

Implementation

There have been several attempts to produce a true implementation of the relational database model as originally defined by Codd and explained by Date, Darwen and others, but none have been popular successes so far. As of October 2015 Rel is one of the more recent attempts to do this.

The relational model was the first database model to be described in formal mathematical terms. Hierarchical and network databases existed before relational databases, but their specifications were relatively informal. After the relational model was defined, there were many attempts to compare and contrast the different models, and this led to the emergence of more rigorous descriptions of the earlier models; though the procedural nature of the data manipulation interfaces for hierarchical and network databases limited the scope for formalization.

The relational model was invented by Edgar Codd as a general model of data, and subsequently maintained and developed by Chris Date and Hugh Darwen among others. In The Third Manifesto (first published in 1995) Date and Darwen show how the relational model can accommodate certain desired object-oriented features.

Controversies

Codd himself, some years after publication of his 1970 model, proposed a three-valued logic (True, False, Missing/NULL) version of it to deal with missing information, and in his The Relational Model for Database Management Version 2 (1990) he went a step further with a four-valued logic (True, False, Missing but Applicable, Missing but Inapplicable) version. But these have never been implemented, presumably because of attending complexity. SQL's NULL construct was intended to be part of a three-valued logic system, but fell short of that due to logical errors in the standard and in its implementations.

Topics

The fundamental assumption of the relational model is that all data is represented as mathematical n-ary **relations**, an n-ary relation being a subset of the Cartesian product of n domains. In the mathematical model, reasoning about such data is done in two-valued predicate logic, meaning there are two possible evaluations for each proposition: either true or false (and in particular no third value such as unknown, or not applicable, either of which are often associated with the concept of NULL). Data are operated upon by means of a relational calculus or relational algebra, these being equivalent in expressive power.

The relational model of data permits the database designer to create a consistent, logical representation of information. Consistency is achieved by including declared ***constraints*** in the database design, which is usually referred to as the logical schema. The theory includes a process of database normalization whereby a design with certain desirable properties can be selected from a set of logically equivalent alternatives. The access plans and other implementation and operation details are handled by the DBMS engine, and are not reflected in the logical model. This contrasts with common practice for SQL DBMSs in which performance tuning often requires changes to the logical model.

The basic relational building block is the domain or data type, usually abbreviated nowadays to ***type***. A tuple is an ordered set of ***attribute values***. An attribute is an ordered pair of ***attribute name*** and ***type name***. An attribute value is a specific valid value for the type of the attribute. This can be either a scalar value or a more complex type.

A relation consists of a ***heading*** and a ***body***. A heading is a set of attributes. A body (of an n-ary relation) is a set of n-tuples. The heading of the relation is also the heading of each of its tuples.

A relation is defined as a set of n-tuples. In both mathematics and the relational database model, a set is an unordered collection of unique, non-duplicated items, although some DBMSs impose an order to their data. In mathematics, a tuple has an order, and allows for duplication. E. F. Codd originally defined tuples using this mathematical definition. Later, it was one of E. F. Codd›s great insights that using attribute names instead of an ordering would be so much more convenient (in general) in a computer language based on relations. This insight is still being used today. Though the concept has changed, the name tuple has not. An immediate and important consequence of this distinguishing feature is that in the relational model the Cartesian product becomes commutative.

A table is an accepted visual representation of a relation; a tuple is similar to the concept of a row.

A relvar is a named variable of some specific relation type, to which at all times some relation of that type is assigned, though the relation may contain zero tuples.

The basic principle of the relational model is the Information Principle: all information is represented by data values in relations. In accordance with this Principle, a relational database is a set of relvars and the result of every query is presented as a relation.

The consistency of a relational database is enforced, not by rules built into the applications that use it, but rather by constraints, declared as part of the logical schema and enforced by the DBMS for all applications. In general, constraints are expressed using relational comparison operators, of which just one, "is subset of" (⊆), is theoretically sufficient. In practice, several useful shorthands are expected to be available, of which the most important are candidate key (really, superkey) and foreign key constraints.

Interpretation

To fully appreciate the relational model of data it is essential to understand the intended interpretation of a relation.

The body of a relation is sometimes called its extension. This is because it is to be interpreted as a representation of the extension of some predicate, this being the set of true propositions that can be formed by replacing each free variable in that predicate by a name (a term that designates something).

There is a one-to-one correspondence between the free variables of the predicate and the attribute names of the relation heading. Each tuple of the relation body provides attribute values to instantiate the predicate by substituting each of its free variables. The result is a proposition that is deemed, on account of the appearance of the tuple in the relation body, to be true. Contrariwise, every tuple whose heading conforms to that of the relation, but which does not appear in the body is deemed to be false. This assumption is known as the closed world assumption: it is often violated in practical databases, where the absence of a tuple might mean that the truth of the corresponding proposition is unknown. For example, the absence of the tuple (‹John›, ‹Spanish›) from a table of language skills cannot necessarily be taken as evidence that John does not speak Spanish.

For a formal exposition of these ideas, see the section Set-theoretic Formulation, below.

Application to Databases

A **data type** as used in a typical relational database might be the set of integers, the set of character strings, the set of dates, or the two boolean values true and false, and so on. The corresponding **type names** for these types might be the strings "int, char, date, boolean, etc. It is important to understand, though, that relational theory does not dictate what types are to be supported; indeed, nowadays provisions are expected to be available for user-defined types in addition to the built-in ones provided by the system.

Attribute is the term used in the theory for what is commonly referred to as a **column**. Similarly, **table** is commonly used in place of the theoretical term **relation** (though in SQL the term is by no means synonymous with relation). A table data structure is specified as a list of column definitions, each of which specifies a unique column name and the type of the values that are permitted for that column. An **attribute *value*** is the entry in a specific column and row, such as "John Doe" or "35".

A **tuple** is basically the same thing as a **row**, except in an SQL DBMS, where the column values in a row are ordered. (Tuples are not ordered; instead, each attribute value is identified solely by the **attribute name** and never by its ordinal position within the tuple.) An attribute name might be "name" or "age".

A **relation** is a **table** structure definition (a set of column definitions) along with the data appearing in that structure. The structure definition is the **heading** and the data appearing in it is the **body**, a set of rows. A database **relvar** (relation variable) is commonly known as a **base table**. The heading of its assigned value at any time is as specified in the table declaration and its body is that most recently assigned to it by invoking some **update operator** (typically, INSERT, UPDATE, or DELETE). The heading and body of the table resulting from evaluation of some query are determined by the definitions of the operators used in the expression of that query. (Note that in SQL the heading is not always a set of column definitions as described above, because it is possible for a column to have no name and also for two or more columns to have the same name. Also, the body is not always a set of rows because in SQL it is possible for the same row to appear more than once in the same body.)

SQL and the Relational Model

SQL, initially pushed as the standard language for relational databases, deviates from the relational model in several places. The current ISO SQL standard doesn't mention the relational model or use relational terms or concepts. However, it is possible to create a database conforming to the relational model using SQL if one does not use certain SQL features.

The following deviations from the relational model have been noted in SQL. Note that few database servers implement the entire SQL standard and in particular do not allow some of these deviations. Whereas NULL is ubiquitous, for example, allowing duplicate column names within a table or anonymous columns is uncommon.

Duplicate rows

The same row can appear more than once in an SQL table. The same tuple cannot appear more than once in a relation.

Anonymous columns

A column in an SQL table can be unnamed and thus unable to be referenced in expressions. The relational model requires every attribute to be named and referenceable.

Duplicate column names

Two or more columns of the same SQL table can have the same name and therefore cannot be referenced, on account of the obvious ambiguity. The relational model requires every attribute to be referenceable.

Column order significance

The order of columns in an SQL table is defined and significant, one consequence being that SQL's implementations of Cartesian product and union are both noncommutative. The relational model requires there to be no significance to any ordering of the attributes of a relation.

Views without CHECK OPTION

Updates to a view defined without CHECK OPTION can be accepted but the resulting update to the database does not necessarily have the expressed effect on its target. For example, an invocation of INSERT can be accepted but the inserted rows might not all appear in the view, or an invocation of UPDATE can result in rows disappearing from the view. The relational model requires updates to a view to have the same effect as if the view were a base relvar.

Columnless tables unrecognized

SQL requires every table to have at least one column, but there are two relations of degree zero (of cardinality one and zero) and they are needed to represent extensions of predicates that contain no free variables.

NULL

This special mark can appear instead of a value wherever a value can appear in SQL, in particular in place of a column value in some row. The deviation from the relational model arises from the fact that the implementation of this ad hoc concept in SQL involves the use of three-valued logic, under which the comparison of NULL with itself does not yield true but instead yields the third truth value, unknown; similarly the comparison NULL with something other than itself does not yield false but instead yields unknown. It is because of this behaviour in comparisons that NULL is described as a mark rather than a value. The relational model depends on the law of excluded middle under which anything that is not true is false and anything that is not false is true; it also requires every tuple in a relation body to have a value for every attribute of that relation. This particular deviation is disputed by some if only because E. F. Codd himself eventually advocated the use of special marks and a 4-valued logic, but this was based on his observation that there are two distinct reasons why one might want to use a special mark in place of a value, which led opponents of the use of such logics to discover more distinct reasons and at least as many as 19 have been noted, which would require a 21-valued logic. SQL itself uses NULL for several purposes other than to represent value unknown. For example, the sum of the empty set is NULL, meaning zero, the average of the empty set is NULL, meaning undefined, and NULL appearing in the result of a LEFT JOIN can mean no value because there is no matching row in the right-hand operand.

Relational Operations

Users (or programs) request data from a relational database by sending it a query that is written in a special language, usually a dialect of SQL. Although SQL was originally intended for end-users, it is much more common for SQL queries to be embedded into software that provides an easier user interface. Many Web sites, such as Wikipedia, perform SQL queries when generating pages.

In response to a query, the database returns a result set, which is just a list of rows containing the answers. The simplest query is just to return all the rows from a table, but more often, the rows are filtered in some way to return just the answer wanted.

Often, data from multiple tables are combined into one, by doing a join. Conceptually, this is done by taking all possible combinations of rows (the Cartesian product), and then filtering out everything except the answer. In practice, relational database management systems rewrite (optimize) queries to perform faster, using a variety of techniques.

There are a number of relational operations in addition to join. These include project (the process of eliminating some of the columns), restrict (the process of eliminating some of the rows), union (a way of combining two tables with similar structures), difference (that lists the rows in one table that are not found in the other), intersect (that lists the rows found in both tables), and product (mentioned above, which combines each row of one table with each row of the other). Depending on which other sources you consult, there are a number of other operators – many of which can be defined in terms of those listed above. These include semi-join, outer operators such as outer join and outer union, and various forms of division. Then there are operators to rename columns, and summarizing or aggregating operators, and if you permit relation values as attributes (RVA – relation-valued attribute), then operators such as group and ungroup. The SELECT statement in SQL serves to handle all of these except for the group and ungroup operators.

The flexibility of relational databases allows programmers to write queries that were not anticipated by the database designers. As a result, relational databases can be used by multiple applications in ways the original designers did not foresee, which is especially important for databases that might be used for a long time (perhaps several decades). This has made the idea and implementation of relational databases very popular with businesses.

Database Normalization

Relations are classified based upon the types of anomalies to which they're vulnerable. A database that's in the first normal form is vulnerable to all types of anomalies, while a database that's in the domain/key normal form has no modification anomalies. Normal forms are hierarchical in nature. That is, the lowest level is the first normal form, and the database cannot meet the requirements for higher level normal forms without first having met all the requirements of the lesser normal forms.

Examples

Database

An idealized, very simple example of a description of some relvars (relation variables) and their attributes:

- Customer (**Customer ID**, Tax ID, Name, Address, City, State, Zip, Phone, Email,Sex)
- Order (**Order No**, Customer ID, Invoice No, Date Placed, Date Promised, Terms, Status)
- Order Line (**Order No, Order Line No**, Product Code, Qty)
- Invoice (**Invoice No**, Customer ID, Order No, Date, Status)
- Invoice Line (**Invoice No, Invoice Line No**, Product Code, Qty Shipped)
- Product (**Product Code**, Product Description)

In this design we have six relvars: Customer, Order, Order Line, Invoice, Invoice Line and Product. The bold, underlined attributes are candidate keys. The non-bold, underlined attributes are foreign keys.

Usually one candidate key is chosen to be called the primary key and used in preference over the other candidate keys, which are then called alternate keys.

A candidate key is a unique identifier enforcing that no tuple will be duplicated; this would make the relation into something else, namely a bag, by violating the basic definition of a set. Both foreign keys and superkeys (that includes candidate keys) can be composite, that is, can be composed of several attributes. Below is a tabular depiction of a relation of our example Customer relvar; a relation can be thought of as a value that can be attributed to a relvar.

Customer Relation

Customer ID	Tax ID	Name	Address	[More fields...]
1234567890	555-5512222	Munmun	323 Broadway	...
2223344556	555-5523232	Wile E.	1200 Main Street	...
3334445563	555-5533323	Ekta	871 1st Street	...
4232342432	555-5325523	E. F. Codd	123 It Way	...

If we attempted to insert a new customer with the ID 1234567890, this would violate the design of the relvar since **Customer ID** is a primary key and we already have a customer 1234567890. The DBMS must reject a transaction such as this that would render the database inconsistent by a violation of an integrity constraint.

Foreign keys are integrity constraints enforcing that the value of the attribute set is drawn from a candidate key in another relation. For example, in the Order relation the attribute **Customer ID** is a foreign key. A join is the operation that draws on information from several relations at once. By joining relvars from the example above we could query the database for all of the Customers, Orders, and Invoices. If we only wanted the tuples for a specific customer, we would specify this using a restriction condition.

If we wanted to retrieve all of the Orders for Customer 1234567890, we could query the database to return every row in the Order table with **Customer ID** 1234567890 and join the Order table to the Order Line table based on **Order No**.

There is a flaw in our database design above. The Invoice relvar contains an Order No attribute. So, each tuple in the Invoice relvar will have one Order No, which implies that there is precisely one Order for each Invoice. But in reality an invoice can be created against many orders, or indeed for no particular order. Additionally the Order relvar contains an Invoice No attribute, implying that each Order has a corresponding Invoice. But again this is not always true in the real world. An order is sometimes paid through several invoices, and sometimes paid without an invoice. In other words, there can be many Invoices per Order and many Orders per Invoice. This is a **many-to-many** relationship between Order and Invoice (also called a non-specific relationship). To represent this relationship in the database a new relvar should be introduced whose role is to specify the correspondence between Orders and Invoices:

OrderInvoice (**Order No, Invoice No**)

Now, the Order relvar has a one-to-many relationship to the OrderInvoice table, as does the Invoice relvar. If we want to retrieve every Invoice for a particular Order, we can query for all orders where **Order No** in the Order relation equals the **Order No** in OrderInvoice, and where **Invoice No** in OrderInvoice equals the **Invoice No** in Invoice.

Set-Theoretic Formulation

Basic notions in the relational model are relation names and attribute names. We will represent these as strings such as "Person" and "name" and we will usually use the variables $r, s, t, \ldots$ and a, b, c to range over them. Another basic notion is the set of atomic values that contains values such as numbers and strings.

Our first definition concerns the notion of tuple, which formalizes the notion of row or record in a table:

Tuple

A tuple is a partial function from attribute names to atomic values.

Header

A header is a finite set of attribute names.

Projection

The projection of a tuple A is $t[A] = \{(a, v) : (a, v) \in t, a \in A\}$.

The next definition defines relation that formalizes the contents of a table as it is defined in the relational model.

Relation

A relation is a tuple (H, B) with H, the header, and B, the body, a set of tuples that all have the domain H.

Such a relation closely corresponds to what is usually called the extension of a predicate in first-order logic except that here we identify the places in the predicate with attribute names. Usually in the relational model a database schema is said to consist of a set of relation names, the headers that are associated with these names and the constraints that should hold for every instance of the database schema.

Relation Universe

A relation universe U over a header H is a non-empty set of relations with header H.

Relation Schema

A relation schema (H, C) consists of a header H and a predicate C(R) that is defined for all relations R with header H. A relation satisfies a relation schema (H, C) if it has header H and satisfies C.

Key Constraints and Functional Dependencies

One of the simplest and most important types of relation constraints is the key constraint. It tells us that in every instance of a certain relational schema the tuples can be identified by their values for certain attributes.

Superkey

A superkey is written as a finite set of attribute names.

A superkey K holds in a relation (H, B) if:

- $K \subseteq H$ and
- there exist no two distinct tuples $t_1, t_2 \in B$ such that $t_1[K] = t_2[K]$.

A superkey holds in a relation universe U if it holds in all relations in U.

Theorem: A superkey K holds in a relation universe U over H if and only if $K \subseteq H$ and $K \to H$ holds in U.

Candidate key

A superkey K holds as a candidate key for a relation universe U if it holds as a superkey for U and there is no proper subset of K that also holds as a superkey for U.

Functional dependency

A functional dependency (FD for short) is written as $X \to Y$ for X, Y finite sets of attribute names.

A functional dependency $X \to Y$ holds in a relation (H, B) if:

- $X, Y \subseteq H$ and
- $\forall$ tuples $t_1, t_2 \in B, t_1[X] = t_2[X] \Rightarrow t_1[Y] = t_2[Y]$

A functional dependency $X \to Y$ is trivial under a header H if and only if $Y \subseteq X \subseteq H$.

Trivial functional dependency

A functional dependency is trivial under a header H if it holds in all relation universes over H.

Theorem: An FD $X \to Y$ is trivial under a header H if and only if $Y \subseteq X \subseteq H$.

Closure

Armstrong's axioms: The closure of a set of FDs S under a header H, written as S^+, is the smallest superset of S such that:

- $Y \subseteq X \subseteq H \Rightarrow X \to Y \in S^+$ (reflexivity)
- $X \to Y \in S^+ \wedge Y \to Z \in S^+ \Rightarrow X \to Z \in S^+$ (transitivity) and
- $X \to Y \in S^+ \wedge Z \subseteq H \Rightarrow (X \cup Z) \to (Y \cup Z) \in S^+$ (augmentation)

Theorem: Armstrong's axioms are sound and complete; given a header H and a set S of FDs that only contain subsets of H, $X \to Y \in S^+$ if and only if $X \to Y$ holds in all relation universes over H in which all FDs in S hold.

Completion

The completion of a finite set of attributes X under a finite set of FDs S, written as X^+, is the smallest superset of X such that:

- $Y \to Z \in S \wedge Y \subseteq X^+ \Rightarrow Z \subseteq X^+$

The completion of an attribute set can be used to compute if a certain dependency is in the closure of a set of FDs.

Theorem: Given a set S of FDs, $X \to Y \in S^+$ if and only if $Y \subseteq X^+$.

Irreducible Cover

An irreducible cover of a set S of FDs is a set T of FDs such that:

- $S^+ = T^+$
- there exists no $U \subset T$ such that $S^+ = U^+$
- $X \to Y \in T \Rightarrow Y$ is a singleton set and
- $X \to Y \in T \wedge Z \subset X \Rightarrow Z \to Y \notin S^+$.

Algorithm to Derive Candidate Keys from Functional Dependencies

```
INPUT: a set S of FDs that contain only subsets of a header H
OUTPUT: the set C of superkeys that hold as candidate keys in
        all relation universes over H in which all FDs in S hold
begin
  C := ∅;          // found candidate keys
  Q := { H };      // superkeys that contain candidate keys
  while Q <> ∅ do
    let K be some element from Q;
    Q := Q – { K };
    minimal := true;
    for each X->Y in S do
      K' := (K – Y) ∪ X;   // derive new superkey
      if K' ⊂ K then
        minimal := false;
        Q := Q ∪ { K' };
      end if
    end for
    if minimal and there is not a subset of K in C then
      remove all supersets of K from C;
      C := C ∪ { K };
    end if
  end while
end
```

RELATIONAL ALGEBRA

Relational algebra, first created by E.F. Codd while at IBM, is a family of algebras with a well-founded semantics used for modelling the data stored in relational databases, and defining queries on it.

The main application of relational algebra is providing a theoretical foundation for relational databases, particularly query languages for such databases, chief among which is SQL.

Relational algebra received little attention outside of pure mathematics until the publication of E.F. Codd›s relational model of data in 1970. Codd proposed such an algebra as a basis for database query languages.

Five primitive operators of Codd›s algebra are the selection, the projection, the Cartesian product (also called the cross product or cross join), the set union, and the set difference.

Set Operators

The relational algebra uses set union, set difference, and Cartesian product from set theory, but adds additional constraints to these operators.

For set union and set difference, the two relations involved must be union-compatible—that is, the two relations must have the same set of attributes. Because set intersection is defined in terms of set union and set difference, the two relations involved in set intersection must also be union-compatible.

For the Cartesian product to be defined, the two relations involved must have disjoint headers—that is, they must not have a common attribute name.

In addition, the Cartesian product is defined differently from the one in set theory in the sense that tuples are considered to be "shallow" for the purposes of the operation. That is, the Cartesian product of a set of n-tuples with a set of m-tuples yields a set of "flattened" (n + m)-tuples (whereas basic set theory would have prescribed a set of 2-tuples, each containing an n-tuple and an m-tuple). More formally, R × S is defined as follows:

$$R \times S := \{(r_1, r_2, \ldots, r_n, s_1, s_2, \ldots, s_m) | (r_1, r_2, \ldots, r_n) \in R, (s_1, s_2, \ldots, s_m) \in S\}$$

The cardinality of the Cartesian product is the product of the cardinalities of its factors, that is, |R × S| = |R| × |S|.

Projection (π)

A **projection** is a unary operation written as $\Pi_{a_1,\ldots,a_n}(R)$ where $a_1, \ldots, a_n$ is a set of attribute names. The result of such projection is defined as the set that is obtained when all tuples in R are restricted to the set $\{a_1, \ldots, a_n\}$.

Selection (σ)

A **generalized selection** is a unary operation written as $\sigma_\varphi(R)$ where φ is a propositional formula that consists of atoms as allowed in the normal selection and the logical operators $\wedge$ (and), $\vee$ (or) and $\neg$ (negation). This selection selects all those tuples in R for which φ holds.

To obtain a listing of all friends or business associates in an address book, the selection might be written as $\sigma_{\text{isFriend = true} \vee \text{isBusinessContact = true}}(\text{addressBook})$. The result would be a relation containing every attribute of every unique record where isFriend is true or where isBusinessContact is true.

In Codd's 1970 paper, selection is called restriction.

Rename (ρ)

A **rename** is a unary operation written as $\rho_{a/b}(R)$ where the result is identical to R except that the b attribute in all tuples is renamed to an a attribute. This is simply used to rename the attribute of a relation or the relation itself.

To rename the ‹isFriend› attribute to ‹isBusinessContact› in a relation, $\rho_{\text{isBusinessContact / isFriend}}(\text{addressBook})$ might be used.

Joins and Join-Like Operators

Natural Join (⋈)

Natural join (⋈) is a binary operator that is written as (R ⋈ S) where R and S are relations. The result of the natural join is the set of all combinations of tuples in R and S that are equal on their common attribute names. For an example consider the tables Employee and Dept and their natural join:

Employee

Name	EmpId	DeptName
Harry	3415	Finance
Sally	2241	Sales
George	3401	Finance
Harriet	2202	Sales

Dept

DeptName	Manager
Finance	George
Sales	Harriet
Production	Charles

Employee ⋈ Dept

Name	EmpId	DeptName	Manager
Harry	3415	Finance	George
Sally	2241	Sales	Harriet
George	3401	Finance	George
Harriet	2202	Sales	Harriet

This can also be used to define composition of relations. For example, the composition of Employee and Dept is their join as shown above, projected on all but the common attribute DeptName. In category theory, the join is precisely the fiber product.

The natural join is arguably one of the most important operators since it is the relational counterpart of logical AND. Note carefully that if the same variable appears in each of two predicates that are connected by AND, then that variable stands for the same thing and both appearances must always be substituted by the same value. In particular, natural join allows the combination of relations that are associated by a foreign key. For example, in the above example a foreign key probably holds from Employee.DeptName to Dept.DeptName and then the natural join of Employee and Dept combines all employees with their departments. Note that this works because the foreign key holds between attributes with the same name. If this is not the case such as in the foreign key from Dept.manager to Employee.Name then we have to rename these columns before we take the natural join. Such a join is sometimes also referred to as an **equijoin** (see θ-join).

More formally the semantics of the natural join are defined as follows:

$$R \bowtie S = \{r \cup s \mid r \in R \land s \in S \land Fun(r \cup s)\} \qquad (1)$$

where Fun is a predicate that is true for a relation t (in the mathematical sense) iff t is a function. It is usually required that R and S must have at least one common attribute, but if this constraint is omitted, and R and S have no common attributes, then the natural join becomes exactly the Cartesian product.

The natural join can be simulated with Codd›s primitives as follows. Assume that $c_1,...,c_m$ are the attribute names common to R and S, $r_1,...,r_n$ are the attribute names unique to R and $s_1,...,s_k$ are the attribute unique to S. Furthermore, assume that the attribute names $x_1,...,x_m$ are neither in R nor in S. In a first step we can now rename the common attribute names in S:

$$T = \rho_{x_1/c_1,\ldots,x_m/c_m}(S) = \rho_{x_1/c_1}(\rho_{x_2/c_2}(\ldots\rho_{x_m/c_m}(S)\ldots)) \quad (2)$$

Then we take the Cartesian product and select the tuples that are to be joined:

$$P = \sigma_{c_1=x_1,\ldots,c_m=x_m}(R \times T) = \sigma_{c_1=x_1}(\sigma_{c_2=x_2}(\ldots\sigma_{c_m=x_m}(R \times T)\ldots))$$

(3)

Finally we take a projection to get rid of the renamed attributes:

$$U = \pi_{r_1,\ldots,r_n,c_1,\ldots,c_m,s_1,\ldots,s_k}(P) \quad (4)$$

θ-Join and Equijoin

Consider tables Car and Boat which list models of cars and boats and their respective prices. Suppose a customer wants to buy a car and a boat, but she does not want to spend more money for the boat than for the car. The θ-join ($\bowtie_\theta$) on the relationCarPrice ≥ BoatPrice produces a table with all the possible options. When using a condition where the attributes are equal, for example Price, then the condition may be specified as Price=Price or alternatively (Price) itself.

Car

CarModel	CarPrice
CarA	20,000
CarB	30,000
CarC	50,000

Boat

BoatModel	BoatPrice
Boat1	10,000
Boat2	40,000
Boat3	60,000

$Car \bowtie_{CarPrice \geq BoatPrice} Boat$

CarModel	CarPrice	BoatModel	BoatPrice
CarA	20,000	Boat1	10,000
CarB	30,000	Boat1	10,000
CarC	50,000	Boat1	10,000
CarC	50,000	Boat2	40,000

If we want to combine tuples from two relations where the combination condition is not simply the equality of shared attributes then it is convenient to have a more general form of join operator, which is the θ-join (or theta-join). The θ-join is a binary operator that is written as $R \bowtie_{a\theta b} S$ or $R \bowtie_{a\theta v} S$ where a and b are attribute names, θ is a binary relational operator in the set {<, ≤, =, >, ≥}, v is a value constant, and R and S are relations. The result of this operation consists of all combinations of tuples in R and

S that satisfy θ. The result of the θ-join is defined only if the headers of S and R are disjoint, that is, do not contain a common attribute.

The simulation of this operation in the fundamental operations is therefore as follows:

$R \bowtie_\theta S = \sigma_\theta(R \times S)$

In case the operator θ is the equality operator (=) then this join is also called an **equijoin**.

Note, however, that a computer language that supports the natural join and rename operators does not need θ-join as well, as this can be achieved by selection from the result of a natural join (which degenerates to Cartesian product when there are no shared attributes).

Semijoin (⋉)(⋊)

The left semijoin is joining similar to the natural join and written as R ⋉ S where R and S are relations. The result of this semijoin is the set of all tuples in R for which there is a tuple in S that is equal on their common attribute names. For an example consider the tables Employee and Dept and their semi join:

Employee

Name	EmpId	DeptName
Harry	3415	Finance
Sally	2241	Sales
George	3401	Finance
Harriet	2202	Production

Dept

DeptName	Manager
Sales	Sally
Production	Harriet

Employee ⋉ Dept

Name	EmpId	DeptName
Sally	2241	Sales
Harriet	2202	Production

More formally the semantics of the semijoin can be defined as follows:

$R \ltimes S = \{ t : t \in R \wedge \exists s \in S(\text{Fun}(t \cup s)) \}$

where Fun(r) is as in the definition of natural join.

The semijoin can be simulated using the natural join as follows. If $a_1, \ldots, a_n$ are the attribute names of R, then

$R \ltimes S = \pi_{a1,..,an}(R \bowtie S)$.

Since we can simulate the natural join with the basic operators it follows that this also holds for the semijoin.

Antijoin (▷)

The antijoin, written as R ▷ S where R and S are relations, is similar to the semijoin, but the result of an antijoin is only those tuples in R for which there is no tuple in S that is equal on their common attribute names.

For an example consider the tables Employee and Dept and their antijoin:

Employee

Name	EmpId	DeptName
Harry	3415	Finance
Sally	2241	Sales
George	3401	Finance
Harriet	2202	Production

Dept

DeptName	Manager
Sales	Sally
Production	Harriet

Employee ▷ Dept

Name	EmpId	DeptName
Harry	3415	Finance
George	3401	Finance

The antijoin is formally defined as follows:

$R \triangleright S = \{ t : t \in R \wedge \neg\exists s \in S(\text{Fun}\ (t \cup s)) \}$

or

$R \triangleright S = \{ t : t \in R$, there is no tuple s of S that satisfies Fun $(t \cup s) \}$

where Fun $(t \cup s)$ is as in the definition of natural join.

The antijoin can also be defined as the complement of the semijoin, as follows:

$$R \triangleright S = R - R \ltimes S \qquad (5)$$

Given this, the antijoin is sometimes called the anti-semijoin, and the antijoin operator is sometimes written as semijoin symbol with a bar above it, instead of ▷.

Division (÷)

The division is a binary operation that is written as R ÷ S. The result consists of the restrictions of tuples in R to the attribute names unique to R, i.e., in the header of R but not in the header of S, for which it holds that all their combinations with tuples in S are present in R. For an example see the tables Completed, DBProject and their division:

Completed

Student	Task
Fred	Database1
Fred	Database2
Fred	Compiler1
Eugene	Database1
Eugene	Compiler1
Sarah	Database1
Sarah	Database2

DBProject

Task
Database1
Database2

Completed ÷ DBProject

Student
Fred
Sarah

If DBProject contains all the tasks of the Database project, then the result of the division above contains exactly the students who have completed both of the tasks in the Database project. More formally the semantics of the division is defined as follows:

$$R \div S = \{ t[a_1,...,a_n] : t \in R \wedge \forall s \in S ((t[a_1,...,a_n] \cup s) \in R) \} \tag{6}$$

where $\{a_1,...,a_n\}$ is the set of attribute names unique to R and $t[a_1,...,a_n]$ is the restriction of t to this set. It is usually required that the attribute names in the header of S are a subset of those of R because otherwise the result of the operation will always be empty.

The simulation of the division with the basic operations is as follows. We assume that $a_1,...,a_n$ are the attribute names unique to R and $b_1,...,b_m$ are the attribute names of S. In the first step we project R on its unique attribute names and construct all combinations with tuples in S:

$T := \pi_{a1,...,an}(R) \times S$

In the prior example, T would represent a table such that every Student (because Student is the unique key / attribute of the Completed table) is combined with every given Task. So Eugene, for instance, would have two rows, Eugene -> Database1 and Eugene -> Database2 in T.

EG: First, let's pretend that "Completed" has a third attribute called "grade". It's unwanted baggage here, so we must project it off always. In fact in this step we can drop 'Task' from R as well; the multiply puts it back on.

T := πStudent(R) × S // This gives us every possible desired combination, including those that don't actually exist in R, and excluding others (eg Fred | compiler1, which is not a desired combination)

T

Student	Task
Fred	Database1
Fred	Database2
Eugene	Database1
Eugene	Database2
Sarah	Database1
Sarah	Database2

In the next step we subtract R from T

relation:

U := T − R

Note that in U we have the possible combinations that "could have" been in R, but weren't.

EG: Again with projections - T and R need to have identical attribute names/headers.

U := T - πStudent,Task(R) // This gives us a "what's missing" list.

T

Student	Task
Fred	Database1
Fred	Database2
Eugene	Database1
Eugene	Database2
Sarah	Database1
Sarah	Database2

R a.k.a. *Completed*

Student	Task
Fred	Database1
Fred	Database2
Fred	Compiler1
Eugene	Database1
Eugene	Compiler1
Sarah	Database1
Sarah	Database2

U aka *T - R* aka *whats missing*

Student	Task
Eugene	Database2

So if we now take the projection on the attribute names unique to R then we have the restrictions of the tuples in R for which not all combinations with tuples in S were present in R:

$V := \pi_{a1,\ldots,an}(U)$

EG: Project U down to just the attribute(s) in question (Student)

$V := \pi$Student(R)

V

Student
Eugene

So what remains to be done is take the projection of R on its unique attribute names and subtract those in V:

$W := \pi_{a1,\ldots,an}(R) - V$

EG: $W := \pi$Student(R) - V.

πStudent(*R*)

Student
Fred
Eugene
Sarah

V

Student
Eugene

W aka (πStudent(*R*) - *V*) aka desired result

Student
Fred
Sarah

Common Extensions

In practice the classical relational algebra described above is extended with various operations such as outer joins, aggregate functions and even transitive closure.

Outer Joins

Whereas the result of a join (or inner join) consists of tuples formed by combining matching tuples in the two operands, an outer join contains those tuples and additionally some tuples formed by extending an unmatched tuple in one of the operands by "fill" values for each of the attributes of the other operand. Note that outer joins are not considered part of the classical relational algebra discussed so far.

The operators defined in this section assume the existence of a null value, ω, which we do not define, to be used for the fill values; in practice this corresponds to the NULL in SQL. In order to make subsequent selection operations on the resulting table meaningful, a semantic meaning needs to be assigned to nulls; in Codd›s approach the propositional logic used by the selection is extended to a three-valued logic, although we elide those details in this article.

Three outer join operators are defined: left outer join, right outer join, and full outer join. (The word outer is sometimes omitted.)

Left Outer Join

The left outer join is written as R S where R and S are relations. The result of the left outer join is the set of all combinations of tuples in R and S that are equal on their common attribute names, in addition (loosely speaking) to tuples in R that have no matching tuples in S.

For an example consider the tables Employee and Dept and their left outer join:

Employee

Name	EmpId	DeptName
Harry	3415	Finance
Sally	2241	Sales
George	3401	Finance
Harriet	2202	Sales
Tim	1123	Executive

Dept

DeptName	Manager
Sales	Harriet
Production	Charles

Employee ⟕ Dept

Name	EmpId	DeptName	Manager
Harry	3415	Finance	ω
Sally	2241	Sales	Harriet
George	3401	Finance	ω
Harriet	2202	Sales	Harriet
Tim	1123	Executive	ω

In the resulting relation, tuples in S which have no common values in common attribute names with tuples in R take a null value, ω.

Since there are no tuples in Dept with a DeptName of Finance or Executive, ωs occur in the resulting relation where tuples in Employee have a DeptName of Finance or Executive.

Let r_1, r_2, ..., r_n be the attributes of the relation R and let {(ω, ..., ω)} be the singleton relation on the attributes that are unique to the relation S (those that are not attributes of R). Then the left outer join can be described in terms of the natural join (and hence using basic operators) as follows:

$$(R \bowtie S) \cup ((R - \pi_{r_1, r_2, \ldots, r_n}(R \bowtie S)) \times \{(\omega, \ldots \omega)\})$$

Right Outer Joi

The right outer join behaves almost identically to the left outer join, but the roles of the tables are switched.

The right outer join of relations R and S is written as R and S. The result of the right outer join is the set of all combinations of tuples in R and S that are equal on their common attribute names, in addition to tuples in S that have no matching tuples in R.

For example, consider the tables Employee and Dept and their right outer join:

Employee

Name	EmpId	DeptName
Harry	3415	Finance
Sally	2241	Sales
George	3401	Finance
Harriet	2202	Sales
Tim	1123	Executive

Dept

DeptName	Manager
Sales	Harriet
Production	Charles

Employee ⋊ Dept

Name	EmpId	DeptName	Manager
Sally	2241	Sales	Harriet
Harriet	2202	Sales	Harriet
ω	ω	Production	Charles

In the resulting relation, tuples in R which have no common values in common attribute names with tuples in S take a null value, ω.

Since there are no tuples in Employee with a DeptName of Production, ωs occur in the Name attribute of the resulting relation where tuples in DeptName had tuples of Production.

Let s_1, s_2, ..., s_n be the attributes of the relation S and let {(ω, ..., ω)} be the singleton relation on the attributes that are unique to the relation R (those that are not attributes of S). Then, as with the left outer join, the right outer join can be simulated using the natural join as follows:

$$(R \bowtie S) \cup (\{(\omega, \ldots, \omega)\} \times (S - \pi_{s_1, s_2, \ldots, s_n}(R \bowtie S)))$$

Full Outer Join

The **outer join** or **full outer join** in effect combines the results of the left and right outer joins.

The full outer join is written as R and S where R and S are relations. The result of the full outer join is the set of all combinations of tuples in R and S that are equal on their common attribute names, in addition to tuples in S that have no matching tuples in R and tuples in R that have no matching tuples in S in their common attribute names.

For an example consider the tables Employee and Dept and their full outer join:

Employee

Name	EmpId	DeptName
Harry	3415	Finance
Sally	2241	Sales
George	3401	Finance
Harriet	2202	Sales
Tim	1123	Executive

Dept

DeptName	Manager
Sales	Harriet
Production	Charles

Employee ⟗ Dept

Name	EmpId	DeptName	Manager
Harry	3415	Finance	ω
Sally	2241	Sales	Harriet
George	3401	Finance	ω
Harriet	2202	Sales	Harriet
Tim	1123	Executive	ω
ω	ω	Production	Charles

In the resulting relation, tuples in R which have no common values in common attribute names with tuples in S take a null value, ω. Tuples in S which have no common values in common attribute names with tuples in R also take a null value, ω.

The full outer join can be simulated using the left and right outer joins (and hence the natural join and set union) as follows:

R □ S = (R □ S) ∪ (R □ S)

Operations for Domain Computations

There is nothing in relational algebra introduced so far that would allow computations on the data domains (other than evaluation of propositional expressions involving equality). For example, it is not possible using only the algebra introduced so far to write an expression that would multiply the numbers from two columns, e.g. a unit price with a quantity to obtain a total price. Practical query languages have such facilities, e.g. the SQL SELECT allows arithmetic operations to define new columns in the result SELECT unit_price * quantity AS total_price FROM t, and a similar facility is provided more explicitly by Tutorial D's EXTEND keyword. In database theory, this is called **extended projection**.

Aggregation

Furthermore, computing various functions on a column, like the summing up its elements, is also not possible using the relational algebra introduced so far. There are five aggregate functions that are included with most relational database systems. These operations are Sum, Count, Average, Maximum and Minimum. In relational algebra the aggregation operation over a schema (A_1, A_2, ... A_n) is written as follows:

$$G_1, G_2, \ldots, G_m \; g_{f_1(A_1'), f_2(A_2'), \ldots, f_k(A_k')} (r)$$

where each A_j', $1 \leq j \leq k$, is one of the original attributes A_i, $1 \leq i \leq n$.

The attributes preceding the g are grouping attributes, which function like a "group by" clause in SQL. Then there are an arbitrary number of aggregation functions applied to individual attributes. The operation is applied to an arbitrary relation r. The grouping attributes are optional, and if they are not supplied, the aggregation functions are applied across the entire relation to which the operation is applied.

Let's assume that we have a table named Account with three columns, namely Account_Number, Branch_Name and Balance. We wish to find the maximum balance of each branch. This is accomplished by ${}_{Branch\_Name}G_{Max(Balance)}$(Account). To find the highest balance of all accounts regardless of branch, we could simply write $G_{Max(Balance)}$(Account).

Transitive Closure

Although relational algebra seems powerful enough for most practical purposes, there are some simple and natural operators on relations that cannot be expressed by relational algebra. One of them is the transitive closure of a binary relation. Given a domain D, let binary relation R be a subset of D×D. The transitive closure R^+ of R is the smallest subset of D×D that contains R and satisfies the following condition:

$$\forall x \forall y \forall z \left((x, y) \in R^+ \wedge (y, z) \in R^+ \Rightarrow (x, z) \in R^+ \right)$$

There is no relational algebra expression E(R) taking R as a variable argument that produces R^+. This can be proved using the fact that, given a relational expression E for which it is claimed that $E(R) = R^+$, where R is a variable, we can always find an instance $\boldsymbol{r}$ of R (and a corresponding domain **d**) such that $E(\boldsymbol{r}) \neq \boldsymbol{r}^+$.

SQL however officially supports such fixpoint queries since 1999, and it had vendor-specific extensions in this direction well before that.

Use of Algebraic Properties for Query Optimization

Queries can be represented as a tree, where

- the internal nodes are operators,
- leaves are relations,
- subtrees are subexpressions.

Our primary goal is to transform expression trees into equivalent expression trees, where the average size of the relations yielded by subexpressions in the tree is smaller than it was before the optimization. Our secondary goal is to try to form common subexpressions within a single query, or if there is more than one query being evaluated at the same time, in all of those queries. The rationale behind the second goal is that it is enough to compute common subexpressions once, and the results can be used in all queries that contain that subexpression.

Here we present a set of rules that can be used in such transformations.

Selection

Rules about selection operators play the most important role in query optimization. Selection is an operator that very effectively decreases the number of rows in its operand, so if we manage to move the selections in an expression tree towards the leaves, the internal relations (yielded by subexpressions) will likely shrink.

Basic Selection Properties

Selection is idempotent (multiple applications of the same selection have no additional effect beyond the first one), and commutative (the order selections are applied in has no effect on the eventual result).

- $\sigma_A(R) = \sigma_A\sigma_A(R)$
- $\sigma_A\sigma_B(R) = \sigma_B\sigma_A(R)$

Breaking up Selections with Complex Conditions

A selection whose condition is a conjunction of simpler conditions is equivalent to a sequence of selections with those same individual conditions, and selection whose condition is a disjunction is equivalent to a union of selections. These identities can be used to merge selections so that fewer selections need to be evaluated, or to split them so that the component selections may be moved or optimized separately.

- $\sigma_{A\wedge B}(R) = \sigma_A(\sigma_B(R)) = \sigma_B(\sigma_A(R))$
- $\sigma_{A\vee B}(R) = \sigma_A(R) \cup \sigma_B(R)$

Selection and Cross Product

Cross product is the costliest operator to evaluate. If the input relations have N and M rows, the result will contain NM rows. Therefore, it is very important to do our best to decrease the size of both operands before applying the cross product operator.

This can be effectively done if the cross product is followed by a selection operator, e.g. σ_A (R × P). Considering the definition of join, this is the most likely case. If the cross product is not followed by a selection operator, we can try to push down a selection from higher levels of the expression tree using the other selection rules.

In the above case we break up condition A into conditions B, C and D using the split rules about complex selection conditions, so that $A = B \wedge C \wedge D$ and B only contains attributes from R, C contains attributes only from P and D contains the part of A that contains attributes from both R and P. Note, that B, C or D are possibly empty. Then the following holds:

$$\sigma_A(R \times P) = \sigma_{B\wedge C\wedge D}(R \times P) = \sigma_D(\sigma_B(R) \times \sigma_C(P))$$

Selection and Set Operators

Selection is distributive over the setminus, intersection, and union operators. The following three rules are used to push selection below set operations in the expression tree. Note, that in the setminus and the intersection operators it is possible to apply the selection operator to only one of the operands after the transformation. This can make sense in cases, where one of the operands is small, and the overhead of evaluating the selection operator outweighs the benefits of using a smaller relation as an operand.

- $\sigma_A(R \setminus P) = \sigma_A(R) \setminus \sigma_A(P) = \sigma_A(R) \setminus P$
- $\sigma_A(R \cup P) = \sigma_A(R) \cup \sigma_A(P)$
- $\sigma_A(R \cap P) = \sigma_A(R) \cap \sigma_A(P) = \sigma_A(R) \cap P = R \cap \sigma_A(P)$

Selection and Projection

Selection commutes with projection if and only if the fields referenced in the selection condition are a subset of the fields in the projection. Performing selection before projection may be useful if the operand is a cross product or join. In other cases, if the selection condition is relatively expensive to compute, moving selection outside the projection may reduce the number of tuples which must be tested (since projection may produce fewer tuples due to the elimination of duplicates resulting from omitted fields).

$$\pi_{a_1,\ldots,a_n}(\sigma_A(R)) = \sigma_A(\pi_{a_1,\ldots,a_n}(R)) \text{ where fields in } A \subseteq \{a_1,\ldots,a_n\}$$

Projection

Basic Projection Properties

Projection is idempotent, so that a series of (valid) projections is equivalent to the outermost projection.

$$\pi_{a_1,\ldots,a_n}(R \cup P) = \pi_{a_1,\ldots,a_n}(R) \cup \pi_{a_1,\ldots,a_n}(P).$$

Projection and Set Operators

Projection is distributive over set union.

$$\pi_{a_1,\ldots,a_n}(R \cup P) = \pi_{a_1,\ldots,a_n}(R) \cup \pi_{a_1,\ldots,a_n}(P).$$

Projection does not distribute over intersection and set difference. Counterexamples are given by:

$$\pi_A(\{\langle A = a, B = b\rangle\} \cap \{\langle A = a, B = b'\rangle\}) = \emptyset$$

$$\pi_A(\{\langle A = a, B = b\rangle\}) \cap \pi_A(\{\langle A = a, B = b'\rangle\}) = \{\langle A = a\rangle\}$$

and

$$\pi_A(\{\langle A = a, B = b\rangle\} \setminus \{\langle A = a, B = b'\rangle\}) = \{\langle A = a\rangle\}$$

$$\pi_A(\{\langle A = a, B = b\rangle\}) \setminus \pi_A(\{\langle A = a, B = b'\rangle\}) = \emptyset,$$

where b is assumed to be distinct from b›.

Rename

Basic Rename Properties

Successive renames of a variable can be collapsed into a single rename. Rename operations which have no variables in common can be arbitrarily reordered with respect to one another, which can be exploited to make successive renames adjacent so that they can be collapsed.

- $\rho_{a/b}(\rho_{b/c}(R)) = \rho_{a/c}(R)$
- $\rho_{a/b}(\rho_{c/d}(R)) = \rho_{c/d}(\rho_{a/b}(R))$

Rename and Set Operators

Rename is distributive over set difference, union, and intersection.

- $\rho_{a/b}(R \setminus P) = \rho_{a/b}(R) \setminus \rho_{a/b}(P)$
- $\rho_{a/b}(R \cup P) = \rho_{a/b}(R) \cup \rho_{a/b}(P)$
- $\rho_{a/b}(R \cap P) = \rho_{a/b}(R) \cap \rho_{a/b}(P)$

Product and Union

Cartesian product is distributive over union.

- $(A \times B) \cup (A \times C) = A \times (B \cup C)$

Implementations

The first query language to be based on Codd's algebra was ISBL, and this pioneering work has been acclaimed by many authorities as having shown the way to make Codd›s idea into a useful language. Business System 12 was a short-lived industry-strength relational DBMS that followed the ISBL example.

In 1998 Chris Date and Hugh Darwen proposed a language called **Tutorial D** intended for use in teaching relational database theory, and its query language also draws on ISBL's ideas. Rel is an implementation of Tutorial D.

Even the query language of SQL is loosely based on a relational algebra, though the operands in SQL (tables) are not exactly relations and several useful theorems about the relational algebra do not hold in the SQL counterpart (arguably to the detriment of optimisers and/or users). The SQL table model is a bag (multiset), rather than a set. For example, the expression $(R \cup S) - T = (R - T) \cup (S - T)$ is a theorem for relational algebra on sets, but not for relational algebra on bags; for a treatment of relational algebra on bags of the "Complete" textbook by Garcia-Molina, Ullman and Widom.

RELATIONAL CALCULUS

Relational calculus consists of two calculi, the tuple relational calculus and the domain relational calculus, that are part of the relational model for databases and provide a declarative way to specify database queries. This in contrast to the relational algebra, which is also part of the relational model but provides a more procedural way for specifying queries.

The relational algebra might suggest these steps to retrieve the phone numbers and names of book stores that supply Some Sample Book:

- Join book stores and titles over the BookstoreID.
- Restrict the result of that join to tuples for the book Some Sample Book.
- Project the result of that restriction over StoreName and StorePhone.

The relational calculus would formulate a descriptive, declarative way:

Get StoreName and StorePhone for book stores such that there exists a title BK with the same BookstoreID value and with a BookTitle value of Some Sample Book.

The relational algebra and the relational calculus are essentially logically equivalent: for any algebraic expression, there is an equivalent expression in the calculus, and vice versa. This result is known as Codd's theorem.

REFERENCES

1. Devlin, Keith, 1993. The Joy of Sets (2nd ed.). Springer Verlag, ISBN 0-387-94094-4
2. Ferreirós, Jose, 2007 (1999). Labyrinth of Thought: A history of set theory and its role in modern mathematics. Basel, Birkhäuser. ISBN 978-3-7643-8349-7
3. Johnson, Philip, 1972. A History of Set Theory. Prindle, Weber & Schmidt ISBN 0-87150-154-6
4. Kunen, Kenneth, 1980. Set Theory: An Introduction to Independence Proofs. North-Holland, ISBN 0-444-85401-0.
5. Potter, Michael, 2004. Set Theory and Its Philosophy: A Critical Introduction. Oxford University Press.
6. Tiles, Mary, 2004 (1989). The Philosophy of Set Theory: An Historical Introduction to Cantor's Paradise. Dover Publications. ISBN 978-0-486-43520-6
7. Foreman, Matthew, Akihiro Kanamori, eds. Handbook of Set Theory. 3 vols., 2010. Each chapter surveys some aspect of contemporary research in set theory. Does not cover established elementary set theory, on which see Devlin (1993).
8. Hazewinkel, Michiel, ed. (2001), "Axiomatic set theory", Encyclopedia of Mathematics, Springer, ISBN 978-1-55608-010-4
9. Hazewinkel, Michiel, ed. (2001), "Set theory", Encyclopedia of Mathematics, Springer, ISBN 978-1-55608-010-4
10. Jech, Thomas (2002). "Set Theory", Stanford Encyclopedia of Philosophy.
11. Schoenflies, Arthur (1898). Mengenlehre in Klein's encyclopedia.
12. Online books, and library resources in your library and in other libraries about set theory
13. Adámek, Jiří; Herrlich, Horst; Strecker, George E. (2004). Abstract and Concrete Categories. Heldermann Verlag Berlin.
14. Barr, Michael; Wells, Charles (2012), Category Theory for Computing Science, Reprints in Theory and Applications of Categories, 22 (3rd ed.).

15. Barr, Michael; Wells, Charles (2005), Toposes, Triples and Theories, Reprints in Theory and Applications of Categories, 12 (revised ed.), MR 2178101.
16. Borceux, Francis (1994). Handbook of categorical algebra. Encyclopedia of Mathematics and its Applications 50-52. Cambridge University Press.
17. Freyd, Peter J. (1964). Abelian Categories. New York: Harper and Row.
18. Freyd, Peter J.; Scedrov, Andre (1990). Categories, allegories. North Holland Mathematical Library. 39. North Holland. ISBN 978-0-08-088701-2.
19. Goldblatt, Robert (2006) [1979]. Topoi: The Categorial Analysis of Logic. Studies in logic and the foundations of mathematics. 94 (Reprint, revised ed.). Dover Publications. ISBN 978-0-486-45026-1.
20. Herrlich, Horst; Strecker, George E. (2007), Category Theory (3rd ed.), Heldermann Verlag Berlin, ISBN 978-3-88538-001-6.
21. Kashiwara, Masaki; Schapira, Pierre (2006). Categories and Sheaves. Grundlehren der Mathematischen Wissenschaften. 332. Springer. ISBN 978-3-540-27949-5.
22. Lawvere, F. William; Rosebrugh, Robert (2003). Sets for Mathematics. Cambridge University Press. ISBN 978-0-521-01060-3.
23. Lawvere, F. W.; Schanuel, Stephen Hoel (2009) [1997]. Conceptual Mathematics: A First Introduction to Categories (2nd ed.). Cambridge University Press. ISBN 978-0-521-89485-2.
24. Leinster, Tom (2004). Higher Operads, Higher Categories. London Math. Society Lecture Note Series. 298. Cambridge University Press. ISBN 978-0-521-53215-0.
25. Leinster, Tom (2014). Basic Category Theory. Cambridge University Press.
26. Lurie, Jacob (2009). Higher Topos Theory. Annals of Mathematics Studies. 170. Princeton, NJ: Princeton University Press. arXiv:math.CT/0608040Freely accessible. ISBN 978-0-691-14049-0. MR 2522659.
27. Mac Lane, Saunders (1998). Categories for the Working Mathematician. Graduate Texts in Mathematics. 5 (2nd ed.). Springer-Verlag. ISBN 0-387-98403-8. MR 1712872.
28. Mac Lane, Saunders; Birkhoff, Garrett (1999) [1967]. Algebra (2nd ed.). Chelsea. ISBN 0-8218-1646-2.
29. Martini, A.; Ehrig, H.; Nunes, D. (1996). "Elements of basic category theory". Technical Report. Technical University Berlin. 96 (5).
30. May, Peter (1999). A Concise Course in Algebraic Topology. University of Chicago Press. ISBN 0-226-51183-9.
31. Guerino, Mazzola (2002). The Topos of Music, Geometric Logic of Concepts, Theory, and Performance. Birkhäuser. ISBN 3-7643-5731-2.
32. Pedicchio, Maria Cristina; Tholen, Walter, eds. (2004). Categorical foundations. Special topics in order, topology, algebra, and sheaf theory. Encyclopedia of Mathematics and Its Applications. 97. Cambridge: Cambridge University Press. ISBN 0-521-83414-7. Zbl 1034.18001.

33. Pierce, Benjamin C. (1991). Basic Category Theory for Computer Scientists. MIT Press. ISBN 978-0-262-66071-6.

34. Schalk, A.; Simmons, H. (2005). An introduction to Category Theory in four easy movements (PDF). Notes for a course offered as part of the MSc. in Mathematical Logic, Manchester University.

35. Simpson, Carlos. Homotopy theory of higher categories. arXiv:1001.4071Freely accessible., draft of a book.

36. Taylor, Paul (1999). Practical Foundations of Mathematics. Cambridge Studies in Advanced Mathematics. 59. Cambridge University Press. ISBN 978-0-521-63107-5.

37. Turi, Daniele (1996–2001). "Category Theory Lecture Notes" (PDF). Retrieved 11 December 2009. Based on Mac Lane 1998.

38. Jean-Pierre Marquis (2008). From a Geometrical Point of View: A Study of the History and Philosophy of Category Theory. Springer Science & Business Media. ISBN 978-1-4020-9384-5.

39. Codd, E.F (1969), Derivability, Redundancy, and Consistency of Relations Stored in Large Data Banks, Research Report, IBM.

40. Codd, E.F (1970). "A Relational Model of Data for Large Shared Data Banks". Communications of the ACM. Classics. 13 (6): 377–87. doi:10.1145/362384.362685.

41. Data Integration Glossary (PDF), US: Department of Transportation, August 2001.

42. Codd, E. F (1990), The Relational Model for Database Management, Addison-Wesley, pp. 371–388, ISBN 0-201-14192-2.

43. Date, C.J (2006). "18. Why Three- and Four-Valued Logic Don't Work". Date on Database: Writings 2000–2006. Apress. pp. 329–41. ISBN 978-1-59059-746-0.

44. Date, C.J (2004). An Introduction to Database Systems (8 ed.). Addison Wesley. pp. 592–97. ISBN 0-321-19784-4.

45. David M. Kroenke, Database Processing: Fundamentals, Design, and Implementation (1997), Prentice-Hall, Inc., pages 130–144

46. Codd, E.F. (June 1970). "A Relational Model of Data for Large Shared Data Banks". Communications of the ACM. 13 (6): 377–387. doi:10.1145/362384.362685.

47. In Unicode, the bowtie symbol is ⋈ (U+22C8).

48. In Unicode, the ltimes symbol is ⋉ (U+22C9). The rtimes symbol is ⋊ (U+22CA)

49. In Unicode, the Antijoin symbol is ▷ (U+25B7).

50. M. Tamer Özsu; Patrick Valduriez (2011). Principles of Distributed Database Systems (3rd ed.). Springer. p. 46. ISBN 978-1-4419-8833-1.

51. Patrick O›Neil; Elizabeth O›Neil (2001). Database: Principles, Programming, and Performance, Second Edition. Morgan Kaufmann. p. 120. ISBN 978-1-55860-438-4.

52. In Unicode, the Left outer join symbol is (U+27D5).

53. In Unicode, the Right outer join symbol is (U+27D6).

54. In Unicode, the Full Outer join symbol is (U+27D7).
55. C. J. Date (2011). SQL and Relational Theory: How to Write Accurate SQL Code. O'Reilly Media, Inc. pp. 133–135. ISBN 978-1-4493-1974-8.
56. Hector Garcia-Molina; Jeffrey D. Ullman; Jennifer Widom (2009). Database systems: the complete book (2nd ed.). Pearson Prentice Hall. ISBN 978-0-13-187325-4.
57. Aho, Alfred V.; Jeffrey D. Ullman (1979). "Universality of data retrieval languages". Proceedings of the 6th ACM SIGACT-SIGPLAN symposium on Principles of programming languages: 110–119. doi:10.1145/567752.567763.

CHAPTER 3

Combinatorics

Chapter Contents

- **Introduction**
- **Approaches and Subfields of Combinatorics**
- **Related Fields**
- **Combinatorial Game Theory**

Combinatorics is the branch of mathematics studying the enumeration, combination, and permutation of sets of elements and the mathematical relations that characterize their properties.

Mathematicians sometimes use the term "combinatorics" to refer to a larger subset of discrete mathematics that includes graph theory. In that case, what is commonly called combinatorics is then referred to as "enumeration."

INTRODUCTION

Combinatorics is a branch of mathematics concerning the study of finite or countable discrete structures. Aspects of combinatorics include counting the structures of a given kind and size (enumerative combinatorics), deciding when certain criteria can be met, and constructing and analyzing objects meeting the criteria (as in combinatorial designs and matroid theory), finding "largest", "smallest", or "optimal" objects (extremal combinatorics and combinatorial optimization), and studying combinatorial structures arising in an algebraic context, or applying algebraic techniques to combinatorial problems (algebraic combinatorics).

Combinatorial problems arise in many areas of pure mathematics, notably in algebra, probability theory, topology, and geometry, and combinatorics also has many applications in mathematical optimization, computer science, ergodic theory and statistical physics. Many combinatorial questions have historically been considered in isolation, giving an *ad hoc* solution to a problem arising in some mathematical context. In the later twentieth century, however, powerful and general theoretical methods were developed, making combinatorics into an independent branch of mathematics in its own right. One of the oldest and most accessible parts of combinatorics is graph theory, which also has numerous natural connections to other areas. Combinatorics is used frequently in computer science to obtain formulas and estimates in the analysis of algorithms.

A mathematician who studies combinatorics is called a **combinatorialist** or a **combinatorist**.

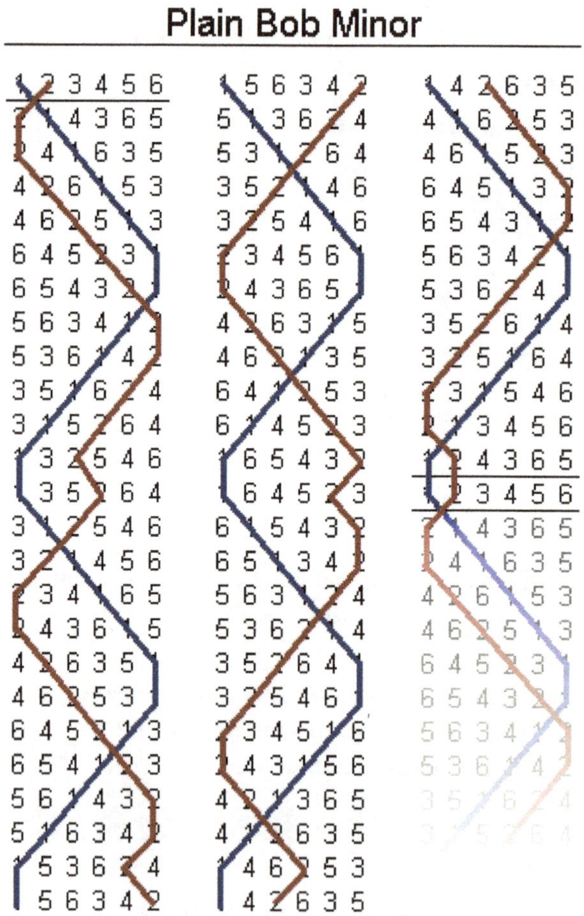

An example of change ringing (with six bells), one of the earliest nontrivial results in Graph Theory.

Basic combinatorial concepts and enumerative results appeared throughout the ancient world. In the 6th century BCE, ancient Indian physician Sushruta asserts in Sushruta

Samhita that 63 combinations can be made out of 6 different tastes, taken one at a time, two at a time, etc., thus computing all $2^6 - 1$ possibilities. Greek historian Plutarch discusses an argument between Chrysippus (3rd century BCE) and Hipparchus (2nd century BCE) of a rather delicate enumerative problem, which was later shown to be related to Schröder–Hipparchus numbers.In the *Ostomachion*, Archimedes (3rd century BCE) considers a tiling puzzle.

In the Middle Ages, combinatorics continued to be studied, largely outside of the European civilization. The Indian mathematician Mahāvīra (c. 850) provided formulae for the number of permutations and combinations, and these formulas may have been familiar to Indian mathematicians as early as the 6th century CE. The philosopher and astronomer Rabbi Abraham ibn Ezra (c. 1140) established the symmetry of binomial coefficients, while a closed formula was obtained later by the talmudist and mathematician Levi ben Gerson (better known as Gersonides), in 1321. The arithmetical triangle— a graphical diagram showing relationships among the binomial coefficients— was presented by mathematicians in treatises dating as far back as the 10th century, and would eventually become known as Pascal›s triangle. Later, in Medieval England, campanology provided examples of what is now known as Hamiltonian cycles in certain Cayley graphs on permutations.

During the Renaissance, together with the rest of mathematics and the sciences, combinatorics enjoyed a rebirth. Works of Pascal, Newton, Jacob Bernoulli and Euler became foundational in the emerging field. In modern times, the works of J. J. Sylvester (late 19th century) and Percy MacMahon (early 20th century) helped lay the foundation for enumerative and algebraic combinatorics. Graph theory also enjoyed an explosion of interest at the same time, especially in connection with the four color problem.

In the second half of the 20th century, combinatorics enjoyed a rapid growth, which led to establishment of dozens of new journals and conferences in the subject. In part, the growth was spurred by new connections and applications to other fields, ranging from algebra to probability, from functional analysis to number theory, etc. These connections shed the boundaries between combinatorics and parts of mathematics and theoretical computer science, but at the same time led to a partial fragmentation of the field.

APPROACHES AND SUBFIELDS OF COMBINATORICS

Enumerative Combinatorics

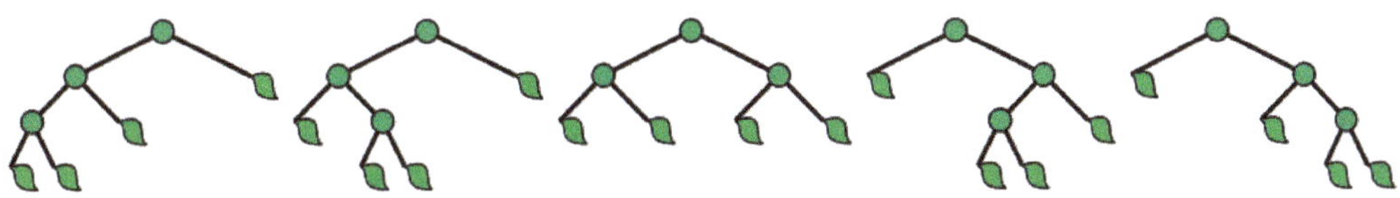

Five binary trees on three vertices, an example of Catalan numbers.

Enumerative combinatorics is the most classical area of combinatorics and concentrates on counting the number of certain combinatorial objects. Although counting the number of elements in a set is a rather broad mathematical problem, many of the problems that arise in applications have a relatively simple combinatorial description. Fibonacci numbers is the basic example of a problem in enumerative combinatorics. The twelvefold way provides a unified framework for counting permutations, combinations and partitions.

Analytic Combinatorics

Analytic combinatorics concerns the enumeration of combinatorial structures using tools from complex analysis and probability theory. In contrast with enumerative combinatorics, which uses explicit combinatorial formulae and generating functions to describe the results, analytic combinatorics aims at obtaining asymptotic formulae.

Partition Theory

A plane partition.

Partition theory studies various enumeration and asymptotic problems related to integer partitions, and is closely related to q-series, special functions and orthogonal polynomials. Originally a part of number theory and analysis, it is now considered a part of combinatorics or an independent field. It incorporates the bijective approach and various tools in analysis and analytic number theory, and has connections with statistical mechanics.

Graph Theory

Graphs are basic objects in combinatorics. The questions range from counting (e.g., the number of graphs on n vertices with k edges) to structural (e.g., which graphs contain Hamiltonian cycles) to algebraic questions (e.g., given a graph G and two numbers x and y, does the Tutte polynomial$T_G(x,y)$ have a combinatorial interpretation?). It should be noted that while there are very strong connections between graph theory and combinatorics, these two are sometimes thought of as separate subjects. This is due to the fact that while combinatorial methods apply to many graph theory problems, the two are generally used to seek solutions to different problems.

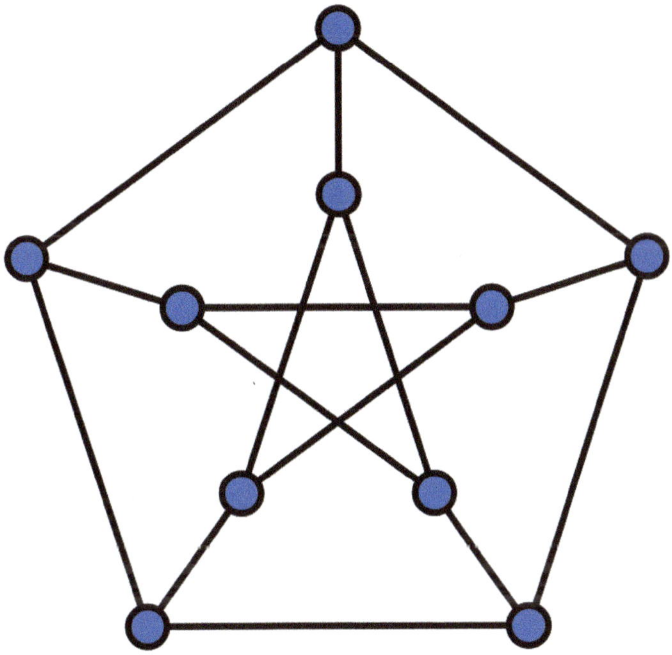

Petersen graph.

Design Theory

Design theory is a study of combinatorial designs, which are collections of subsets with certain intersection properties. Block designs are combinatorial designs of a special type. This area is one of the oldest parts of combinatorics, such as in Kirkman›s schoolgirl problem proposed in 1850. The solution of the problem is a special case of a Steiner system, which systems play an important role in the classification of finite simple groups. The area has further connections to coding theory and geometric combinatorics.

Finite Geometry

Finite geometry is the study of geometric systems having only a finite number of points. Structures analogous to those found in continuous geometries (Euclidean plane, real projective space, etc.) but defined combinatorially are the main items studied. This area provides a rich source of examples for design theory. It should not be confused with discrete geometry (combinatorial geometry).

Order Theory

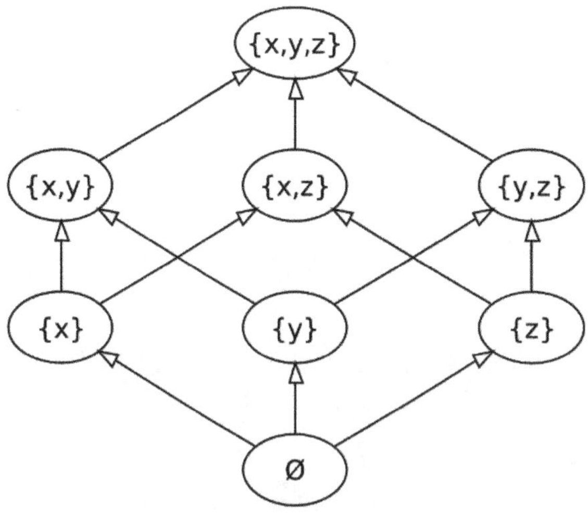

Hasse diagram of the powerset of {x,y,z} ordered by inclusion.

Order theory is the study of partially ordered sets, both finite and infinite. Various examples of partial orders appear in algebra, geometry, number theory and throughout combinatorics and graph theory. Notable classes and examples of partial orders include lattices and Boolean algebras.

Matroid Theory

Matroid theory abstracts part of geometry. It studies the properties of sets (usually, finite sets) of vectors in a vector space that do not depend on the particular coefficients in a linear dependence relation. Not only the structure but also enumerative properties belong to matroid theory. Matroid theory was introduced by Hassler Whitney and studied as a part of order theory. It is now an independent field of study with a number of connections with other parts of combinatorics.

Extremal Combinatorics

Extremal combinatorics studies extremal questions on set systems. The types of questions addressed in this case are about the largest possible graph which satisfies certain properties. For example, the largest triangle-free graph on $2n$ vertices is a complete bipartite graph $K_{n,n}$. Often it is too hard even to find the extremal answer $f(n)$ exactly and one can only give an asymptotic estimate.

Ramsey theory is another part of extremal combinatorics. It states that any sufficiently large configuration will contain some sort of order. It is an advanced generalization of the pigeonhole principle.

Probabilistic Combinatorics

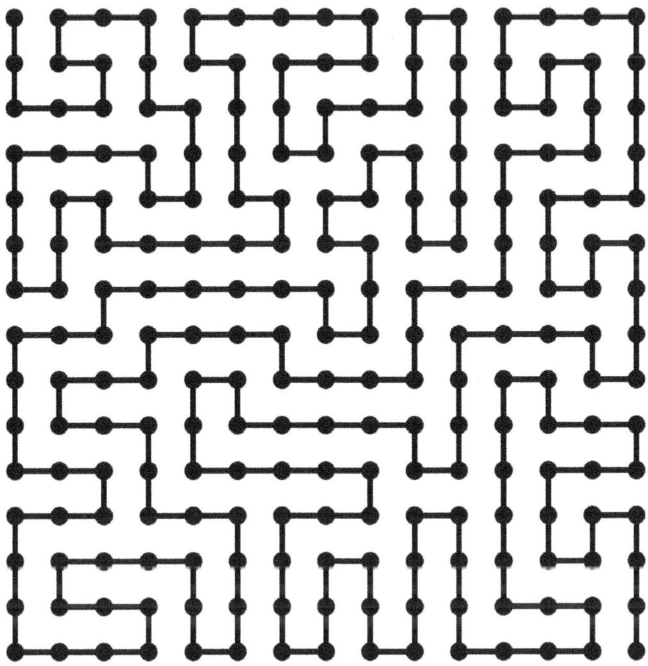

Self-avoiding walk in a square grid graph.

In probabilistic combinatorics, the questions are of the following type: what is the probability of a certain property for a random discrete object, such as a random graph? For instance, what is the average number of triangles in a random graph? Probabilistic methods are also used to determine the existence of combinatorial objects with certain prescribed properties (for which explicit examples might be difficult to find), simply by observing that the probability of randomly selecting an object with those properties is greater than 0. This approach (often referred to as *the*probabilistic method) proved highly effective in applications to extremal combinatorics and graph theory. A closely related area is the study of finite Markov chains, especially on combinatorial objects. Here again probabilistic tools are used to estimate the mixing time.

Often associated with Paul Erdős, who did the pioneering work on the subject, probabilistic combinatorics was traditionally viewed as a set of tools to study problems in other parts of combinatorics. However, with the growth of applications to analysis of algorithms in computer science, as well as classical probability, additive and probabilistic number theory, the area recently grew to become an independent field of combinatorics.

Algebraic Combinatorics

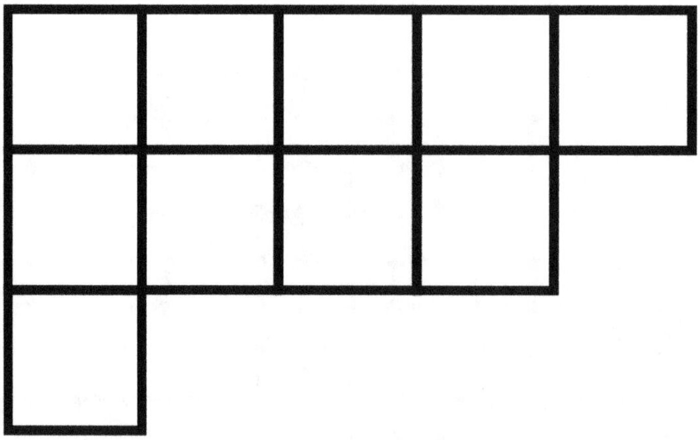

Young diagram of a partition (5,4,1).

Algebraic combinatorics is an area of mathematics that employs methods of abstract algebra, notably group theory and representation theory, in various combinatorial contexts and, conversely, applies combinatorial techniques to problems in algebra. Algebraic combinatorics is continuously expanding its scope, in both topics and techniques, and can be seen as the area of mathematics where the interaction of combinatorial and algebraic methods is particularly strong and significant.

Combinatorics on Words

01101001
10010110

Construction of a Thue–Morse infinite word.

Combinatorics on words deals with formal languages. It arose independently within several branches of mathematics, including number theory, group theory and probability. It has applications to enumerative combinatorics, fractal analysis, theoretical computer science, automata theory and linguistics. While many applications are new, the classical Chomsky–Schützenberger hierarchy of classes of formal grammars is perhaps the best-known result in the field.

Geometric Combinatorics

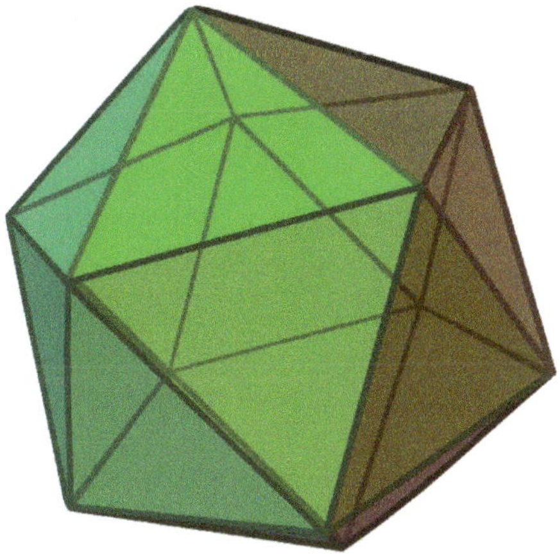

An icosahedron.

Geometric combinatorics is related to convex and discrete geometry, in particular polyhedral combinatorics. It asks, for example, how many faces of each dimension a convex polytope can have. Metric properties of polytopes play an important role as well, e.g. the Cauchy theorem on the rigidity of convex polytopes. Special polytopes are also considered, such as permutohedra, associahedra and Birkhoff polytopes. We should note that combinatorial geometry is an old fashioned name for discrete geometry.

Topological Combinatorics

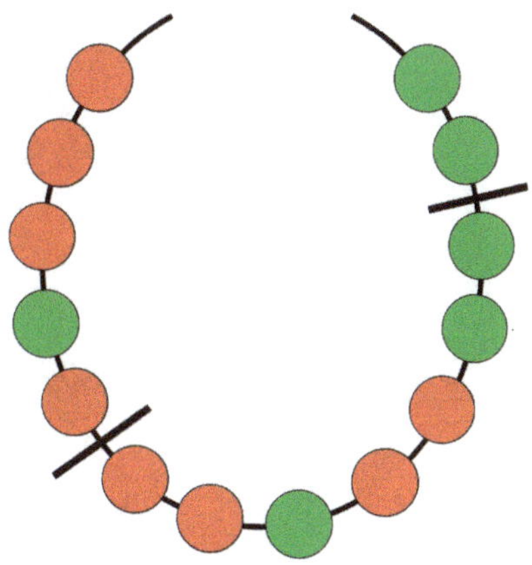

Splitting a necklace with two cuts.

Combinatorial analogs of concepts and methods in topology are used to study graph coloring, fair division, partitions, partially ordered sets, decision trees, necklace problems and discrete Morse theory. It should not be confused with combinatorial topology which is an older name for algebraic topology.

Arithmetic Combinatorics

Arithmetic combinatorics arose out of the interplay between number theory, combinatorics, ergodic theory and harmonic analysis. It is about combinatorial estimates associated with arithmetic operations (addition, subtraction, multiplication, and division). *Additive combinatorics* refers to the special case when only the operations of addition and subtraction are involved. One important technique in arithmetic combinatorics is the ergodic theory of dynamical systems.

Infinitary Combinatorics

Infinitary combinatorics, or combinatorial set theory, is an extension of ideas in combinatorics to infinite sets. It is a part of set theory, an area of mathematical logic, but uses tools and ideas from both set theory and extremal combinatorics.

Gian-Carlo Rota used the name *continuous combinatorics* to describe geometric probability, since there are many analogies between *counting* and *measure*.

RELATED FIELDS

Kissing spheres are connected to both coding theory and discrete geometry.

Combinatorial Optimization

Combinatorial optimization is the study of optimization on discrete and combinatorial objects. It started as a part of combinatorics and graph theory, but is now viewed as a branch of applied mathematics and computer science, related to operations research, algorithm theory and computational complexity theory.

Coding Theory

Coding theory started as a part of design theory with early combinatorial constructions of error-correcting codes. The main idea of the subject is to design efficient and reliable methods of data transmission. It is now a large field of study, part of information theory.

Discrete and Computational Geometry

Discrete geometry (also called combinatorial geometry) also began as a part of combinatorics, with early results on convex polytopes and kissing numbers. With the emergence of applications of discrete geometry to computational geometry, these two fields partially merged and became a separate field of study. There remain many connections with geometric and topological combinatorics, which themselves can be viewed as outgrowths of the early discrete geometry.

Combinatorics and Dynamical Systems

Combinatorial aspects of dynamical systems is another emerging field. Here dynamical systems can be defined on combinatorial objects. See for example graph dynamical system.

Combinatorics and Physics

There are increasing interactions between combinatorics and physics, particularly statistical physics. Examples include an exact solution of the Ising model, and a connection between the Potts model on one hand, and the chromatic and Tutte polynomials on the other hand.

COMBINATORIAL GAME THEORY

Combinatorial game theory (CGT) is a branch of mathematics and theoretical computer science that typically studies sequential games with perfect information. Study has been largely confined to two-player games that have a *position* in which the players take turns changing in defined ways or *moves* to achieve a defined winning condition. CGT has not traditionally studied games of chance or those that use imperfect or incomplete information, favoring games that offer perfect information in which the state of the game and the set of available moves is always known by both players. However, as mathematical techniques advance, the types of game that can be mathematically analyzed expands, thus the boundaries of the field are ever changing. Scholars will generally define what they mean by a "game" at the beginning of a paper, and these definitions often vary as they are specific to the game being analyzed and are not meant to represent the entire scope of the field.

Mathematicians playing Konane at a combinatorial game theory workshop

Combinatorial games include well-known games such as Chess, Checkers, and Go, which are categorized as non-trivial, and Tic-tac-toe, which is categorized as trivial in the sense of being «easy to solve». In CGT, the moves in these and other games are represented as a game tree.

Combinatorial games also include one-player combinatorial puzzles such as Sudoku, and no-player automata, such as Conway›s Game of Life, (although in the strictest definition, «games» can be said to require more than one participant, thus the designations of «puzzle» and «automata».)

Game theory in general includes games of chance, games of imperfect knowledge, and games in which players can move simultaneously, and they tend to represent real-life decision making situations.

CGT has a different emphasis than «traditional» or «economic» game theory, which was initially developed to study games with simple combinatorial structure, but with elements of chance (although it also considers sequential moves, see extensive-form game). Essentially, CGT has contributed new methods for analyzing game trees, for example using surreal numbers, which are a subclass of all two-player perfect-information games. The type of games studied by CGT is also of interest in artificial intelligence, particularly for automated planning and scheduling. In CGT there has been less emphasis on refining practical search algorithms (such as the alpha-beta pruning heuristic included in most artificial intelligence textbooks), but more emphasis on descriptive theoretical results (such as measures of game complexity or proofs of optimal solution existence without necessarily specifying an algorithm, such as the strategy-stealing argument).

An important notion in CGT is that of the solved game. For example, tic-tac-toe is considered a solved game, as it can be proven that any game will result in a draw if both players play optimally. Deriving similar results for games with rich combinatorial structures is difficult. For instance, in 2007 it was announced that checkers has been weakly solved—optimal play by both sides also leads to a draw—but this result was a computer-assisted proof. Other real world games are mostly too complicated to allow complete analysis today, although the theory has had some recent successes in analyzing Go endgames. Applying CGT to a *position* attempts to determine the optimum sequence of moves for both players until the game ends, and by doing so discover the optimum move in any position. In practice, this process is torturously difficult unless the game is very simple.

It can be helpful to distinguish between combinatorial «mathgames» of interest primarily to mathematicians and scientists to ponder and solve, and combinatorial «playgames» of interest to the general population as a form of entertainment and competition. However, a number of games fall into both categories. Nim, for instance, is a playgame instrumental in the foundation of CGT, and one of the first computerized games. Tic-tac-toe is still used to teach basic principles of game AI design to computer science students.

CGT arose in relation to the theory of impartial games, in which any play available to one player must be available to the other as well. One very important such game is nim, which can be solved completely. Nim is an impartial game for two players, and subject to the *normal play condition*, which means that a player who cannot move loses. In the 1930s, the Sprague-Grundy theorem showed that all impartial games are equivalent to heaps in nim, thus showing that major unifications are possible in games considered at a combinatorial level, in which detailed strategies matter, not just pay-offs.

In the 1960s, Elwyn R. Berlekamp, John H. Conway and Richard K. Guy jointly introduced the theory of a partisan game, in which the requirement that a play available to one player be available to both is relaxed. Their results were published in their book *Winning Ways for your Mathematical Plays* in 1982. However, the first work published on the subject was Conway's 1976 book *On Numbers and Games*, also known as ONAG, which introduced the concept of surreal numbers and the generalization to games. *On Numbers and Games* was also a fruit of the collaboration between Berlekamp, Conway, and Guy.

Combinatorial games are generally, by convention, put into a form where one player wins when the other has no moves remaining. It is easy to convert any finite game with only two possible results into an equivalent one where this convention applies. One of the most important concepts in the theory of combinatorial games is that of the sum of two games, which is a game where each player may choose to move either in one game or the other at any point in the game, and a player wins when his opponent has no move in either game. This way of combining games leads to a rich and powerful mathematical structure.

John Conway states in ONAG that the inspiration for the theory of partisan games was based on his observation of the play in go endgames, which can often be decomposed into sums of simpler endgames isolated from each other in different parts of the board.

Examples

The introductory text *Winning Ways* introduced a large number of games, but the following were used as motivating examples for the introductory theory:

- Blue-Red Hackenbush - At the finite level, this partisan combinatorial game allows constructions of games whose values are dyadic rational numbers. At the infinite level, it allows one to construct all real values, as well as many infinite ones that fall within the class of surreal numbers.
- Blue-Red-Green Hackenbush - Allows for additional game values that are not numbers in the traditional sense, for example, star.
- Toads and Frogs - Allows various game values. Unlike most other games, a position is easily represented by a short string of characters.
- Domineering - Various interesting games, such as hot games, appear as positions in Domineering, because there is sometimes an incentive to move, and sometimes not. This allows discussion of a game›s temperature.
- Nim - An impartial game. This allows for the construction of the nimbers. (It can also be seen as a green-only special case of Blue-Red-Green Hackenbush.)

The classic game Go was influential on the early combinatorial game theory, and Berlekamp and Wolfe subsequently developed an endgame and *temperature* theory for it (see references). Armed with this they were able to construct plausible Go endgame positions from which they could give expert Go players a choice of sides and then defeat them either way.

A game, in its simplest terms, is a list of possible «moves» that two players, called *left* and *right*, can make. The game position resulting from any move can be considered to be another game. This idea of viewing games in terms of their possible moves to other games leads to a recursive mathematical definition of games that is standard in combinatorial game theory. In this definition, each game has the notation $\{L|R\}$. L is the set of game positions that the left player can move to, and R is the set of game positions that the right player can move to; each position in L and R is defined as a game using the same notation.

Using Domineering as an example, label each of the sixteen boxes of the four-by-four board by *A1* for the upper leftmost square, *C2* for the third box from the left on the second row from the top, and so on. We use e.g. (D3, D4) to stand for the game position in which a vertical domino has been placed in the bottom right corner. Then, the initial position can be described in combinatorial game theory notation as

$$\{(A1, A2), (B1, B2), \ldots | (A1, B1), (A2, B2), \ldots\}.$$

In standard Cross-Cram play, the players alternate turns, but this alternation is handled implicitly by the definitions of combinatorial game theory rather than being encoded within the game states.

$\{(A1, A2)|(A1, B1)\} = \{\{|\}|\{|\}\}$.

The above game describes a scenario in which there is only one move left for either player, and if either player makes that move, that player wins. (An irrelevant open square at C3 has been omitted from the diagram.) The {|} in each player's move list (corresponding to the single leftover square after the move) is called the zero game, and can actually be abbreviated 0. In the zero game, neither player has any valid moves; thus, the player whose turn it is when the zero game comes up automatically loses.

The type of game in the diagram above also has a simple name; it is called the star game, which can also be abbreviated *. In the star game, the only valid move leads to the zero game, which means that whoever's turn comes up during the star game automatically wins.

An additional type of game, not found in Domineering, is a *loopy* game, in which a valid move of either *left* or *right* is a game that can then lead back to the first game. Checkers, for example, becomes loopy when one of the pieces promotes, as then it can cycle endlessly between two or more squares. A game that does not possess such moves is called *loopfree*.

Game Abbreviations

Numbers

Numbers represent the number of free moves, or the move advantage of a particular player. By convention positive numbers represent an advantage for Left, while negative numbers represent an advantage for Right. They are defined recursively with 0 being the base case.

0 = {|}

1 = {0|}, 2 = {1|}, 3 = {2|}

-1 = {|0}, -2 = {|-1}, -3 = {|-2}

The zero game is a loss for the first player.

The sum of number games behaves like the integers, for example 3 + -2 = 1.

Star

Star, written as * or {0|0}, is a first-player win since either player must (if first to move in the game) move to a zero game, and therefore win.

* + * = 0, because the first player must turn one copy of * to a 0, and then the other player will have to turn the other copy of * to a 0 as well; at this point, the first player would lose, since 0 + 0 admits no moves. The game * is neither positive nor negative; it and all other games in which the first player wins (regardless of which side the player is on) are said to be *fuzzy with* or *confused with* 0; symbolically, we write * || 0.

Up

Up, written as ↑, is a position in combinatorial game theory. In standard notation, ↑ = {0|*}.

−↑ = ↓ (*down*)

Up is strictly positive (↑ > 0), but is infinitesimal. Up is defined in Winning Ways for your Mathematical Plays.

Down

Down, written as ↓, is a position in combinatorial game theory. In standard notation, ↓ = {*|0}.

−↓ = ↑ (*up*)

Down is strictly negative (↓ < 0), but is infinitesimal. Down is defined in Winning Ways for your Mathematical Plays.

"Hot" Games

Consider the game {1|-1}. Both moves in this game are an advantage for the player who makes them; so the game is said to be "hot;" it is greater than any number less than -1, less than any number greater than 1, and fuzzy with any number in between. It is written as ±1. It can be added to numbers, or multiplied by positive ones, in the expected fashion; for example, 4 ± 1 = {5|3}.

Nimbers

An impartial game is one where, at every position of the game, the same moves are available to both players. For instance, Nim is impartial, as any set of objects that can be removed by one player can be removed by the other. However, domineering is not impartial, because one player places horizontal dominoes and the other places vertical ones. Likewise Checkers is not impartial, since the players own different colored pieces. For any ordinal number, one can define an impartial game generalizing Nim in which, on each move, either player may replace the number with any smaller ordinal number; the games defined in this way are known as nimbers. The Sprague–Grundy theorem states that every impartial game is equivalent to a nimber.

The «smallest» nimbers - the simplest and least under the usual ordering of the ordinals - are 0 and *.

REFERENCES

1. Björner, Anders; and Stanley, Richard P.; (2010); *A Combinatorial Miscellany*
2. Bóna, Miklós; (2011); *A Walk Through Combinatorics (3rd Edition)*. ISBN 978-981-4335-23-2, ISBN 978-981-4460-00-2(pbk)

3. Graham, Ronald L.; Groetschel, Martin; and Lovász, László; eds. (1996); *Handbook of Combinatorics*, Volumes 1 and 2. Amsterdam, NL, and Cambridge, MA: Elsevier (North-Holland) and MIT Press. ISBN 0-262-07169-X
4. Lindner, Charles C.; and Rodger, Christopher A.; eds. (1997); *Design Theory*, CRC-Press; 1st. edition (October 31, 1997). ISBN 0-8493-3986-3.
5. Riordan, John (1958); *An Introduction to Combinatorial Analysis*, New York, NY: Wiley & Sons (republished)
6. Stanley, Richard P. (1997, 1999); *Enumerative Combinatorics*, Volumes 1 and 2, Cambridge University Press. ISBN 0-521-55309-1, ISBN 0-521-56069-1
7. van Lint, Jacobus H.; and Wilson, Richard M.; (2001); *A Course in Combinatorics*, 2nd Edition, Cambridge University Press. ISBN 0-521-80340-3
8. Albert, Michael H.; Nowakowski, Richard J.; Wolfe, David (2007). Lessons in Play: An Introduction to Combinatorial Game Theory. A K Peters Ltd. ISBN 978-1-56881-277-9.
9. Beck, József (2008). Combinatorial games: tic-tac-toe theory. Cambridge University Press. ISBN 978-0-521-46100-9.
10. Berlekamp, E.; Conway, J. H.; Guy, R. (1982). Winning Ways for your Mathematical Plays: Games in general. Academic Press. ISBN 0-12-091101-9. 2nd ed., A K Peters Ltd (2001–2004), ISBN 1-56881-130-6, ISBN 1-56881-142-X
11. Berlekamp, E.; Conway, J. H.; Guy, R. (1982). Winning Ways for your Mathematical Plays: Games in particular. Academic Press. ISBN 0-12-091102-7. 2nd ed., A K Peters Ltd (2001–2004), ISBN 1-56881-143-8, ISBN 1-56881-144-6.
12. Berlekamp, Elwyn; Wolfe, David (1997). Mathematical Go: Chilling Gets the Last Point. A K Peters Ltd. ISBN 1-56881-032-6.
13. Bewersdorff, Jörg (2004). Luck, Logic and White Lies: The Mathematics of Games. A K Peters Ltd. ISBN 1-56881-210-8. See especially sections 21–26.
14. Conway, John Horton (1976). On Numbers and Games. Academic Press. ISBN 0-12-186350-6. 2nd ed., A K Peters Ltd (2001), ISBN 1-56881-127-6.
15. Robert A. Hearn; Erik D. Demaine (2009). Games, Puzzles, and Computation. A K Peters, Ltd. ISBN 978-1-56881-322-6.

CHAPTER 4

Graph Theory

Chapter Contents

- **Introduction**
- **Definitions**
- **Applications**
- **Graph Drawing**
- **Graph-Theoretic Data Structures**
- **Problems in Graph Theory**

Graph theory is the study of points and lines. In particular, it involves the ways in which sets of points, called vertices, can be connected by lines or arcs, called edges. Graphs in this context differ from the more familiar coordinate plots that portray mathematical relations and functions.

INTRODUCTION

In mathematics **graph theory** is the study of *graphs*, which are mathematical structures used to model pairwise relations between objects. A graph in this context is made up of *vertices*, *nodes*, or *points* which are connected by *edges*, *arcs*, or *lines*. A graph may be *undirected*, meaning that there is no distinction between the two vertices associated with each edge, or its edges may be *directed* from one vertex to another; see Graph (discrete mathematics) for more detailed definitions and for other variations in the types of graph that are commonly considered. Graphs are one of the prime objects of study in discrete mathematics.

Refer to the glossary of graph theory for basic definitions in graph theory.

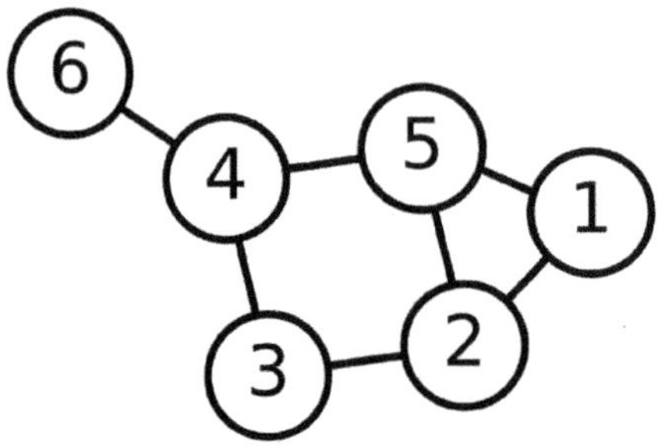

A drawing of a graph

DEFINITIONS

Definitions in graph theory vary. The following are some of the more basic ways of defining graphs and related mathematical structures.

Graph

In the most common sense of the term, a **graph** is an ordered pair $G = (V, E)$ comprising a set V of *vertices* or *nodes* or *points* together with a set E of *edges* or *arcs* or *lines*, which are 2-element subsets of V (i.e. an edge is associated with two vertices, and that association takes the form of the unordered pair comprising those two vertices). To avoid ambiguity, this type of graph may be described precisely as undirected and simple.

Other senses of *graph* stem from different conceptions of the edge set. In one more generalized notion, V is a set together with a relation of *incidence* that associates with each edge two vertices. In another generalized notion, E is a multiset of unordered pairs of (not necessarily distinct) vertices. Many authors call this type of object a multigraph or pseudograph.

All of these variants and others are described more fully below.

The vertices belonging to an edge are called the *ends* or *end vertices* of the edge. A vertex may exist in a graph and not belong to an edge.

V and E are usually taken to be finite, and many of the well-known results are not true (or are rather different) for infinite graphs because many of the arguments fail in the infinite case. The *order* of a graph is $|V|$, its number of vertices. The *size* of a graph is $|E|$, its number of edges. The *degree* or *valency* of a vertex is the number of edges that connect to it, where an edge that connects a vertex to itself (a loop) is counted twice.

For an edge $\{x, y\}$, graph theorists usually use the somewhat shorter notation xy.

APPLICATIONS

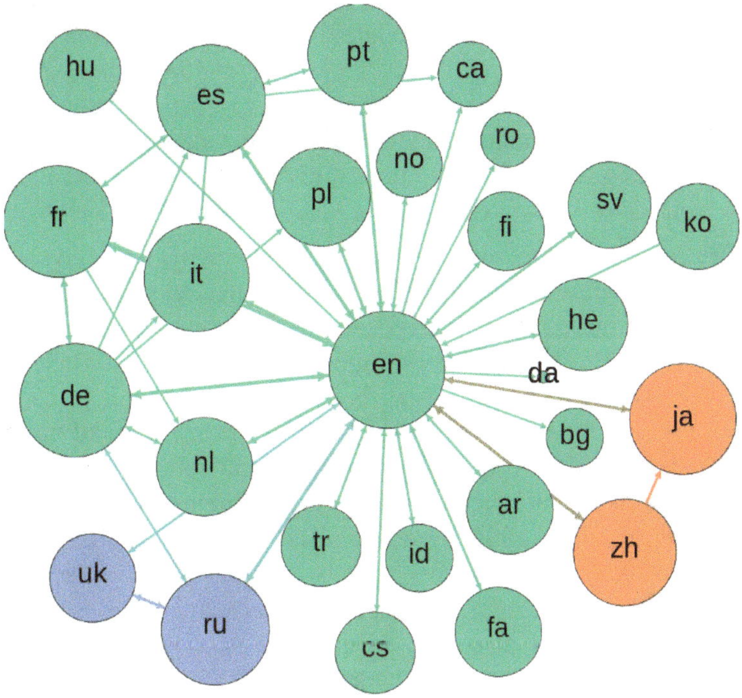

The network graph formed by Wikipedia editors (edges) contributing to different Wikipedia language versions (vertices) during one month in summer 2013

Graphs can be used to model many types of relations and processes in physical, biological, social and information systems. Many practical problems can be represented by graphs. Emphasizing their application to real-world systems, the term *network* is sometimes defined to mean a graph in which attributes (e.g. names) are associated with the nodes and/or edges.

In computer science, graphs are used to represent networks of communication, data organization, computational devices, the flow of computation, etc. For instance, the link structure of a website can be represented by a directed graph, in which the vertices represent web pages and directed edges represent links from one page to another. A similar approach can be taken to problems in social media, travel, biology, computer chip design, and many other fields. The development of algorithms to handle graphs is therefore of major interest in computer science. The transformation of graphs is often formalized and represented by graph rewrite systems. Complementary to graph transformation systems focusing on rule-based in-memory manipulation of graphs are graph databases geared towards transaction-safe, persistent storing and querying of graph-structured data.

Graph-theoretic methods, in various forms, have proven particularly useful in linguistics, since natural language often lends itself well to discrete structure. Traditionally, syntax and compositional semantics follow tree-based structures, whose expressive power lies in the principle of compositionality, modeled in a hierarchical graph. More contemporary approaches such as head-driven phrase structure grammar model the syntax of natural language using typed feature structures, which are directed acyclic graphs. Within lexical semantics, especially as applied to computers, modeling word meaning is easier when a given word is understood in terms of related words; semantic networks are therefore important in computational linguistics. Still other methods in phonology (e.g. optimality theory, which uses lattice graphs) and morphology (e.g. finite-state morphology, using finite-state transducers) are common in the analysis of language as a graph. Indeed, the usefulness of this area of mathematics to linguistics has borne organizations such as TextGraphs, as well as various ‹Net› projects, such as WordNet, VerbNet, and others.

Graph theory is also used to study molecules in chemistry and physics. In condensed matter physics, the three-dimensional structure of complicated simulated atomic structures can be studied quantitatively by gathering statistics on graph-theoretic properties related to the topology of the atoms. In chemistry a graph makes a natural model for a molecule, where vertices represent atoms and edges bonds. This approach is especially used in computer processing of molecular structures, ranging from chemical editors to database searching. In statistical physics, graphs can represent local connections between interacting parts of a system, as well as the dynamics of a physical process on such systems. Similarly, in computational neuroscience graphs can be used to represent functional connections between brain areas that interact to give rise to various cognitive processes, where the vertices represent different areas of the brain and the edges represent the connections between those areas. Graphs are also used to represent the micro-scale channels of porous media, in which the vertices represent the pores and the edges represent the smaller channels connecting the pores.

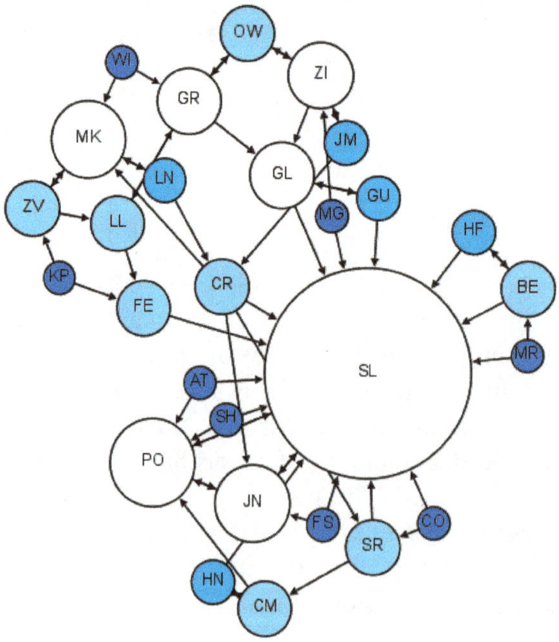

Graph theory in sociology: MorenoSociogram (1953)

Graph theory is also widely used in sociology as a way, for example, to measure actors› prestige or to explore rumor spreading, notably through the use of social network analysis software. Under the umbrella of social networks are many different types of graphs. Acquaintanceship and friendship graphs describe whether people know each other. Influence graphs model whether certain people can influence the behavior of others. Finally, collaboration graphs model whether two people work together in a particular way, such as acting in a movie together.

Likewise, graph theory is useful in biology and conservation efforts where a vertex can represent regions where certain species exist (or inhabit) and the edges represent migration paths, or movement between the regions. This information is important when looking at breeding patterns or tracking the spread of disease, parasites or how changes to the movement can affect other species.

In mathematics, graphs are useful in geometry and certain parts of topology such as knot theory. Algebraic graph theory has close links with group theory.

A graph structure can be extended by assigning a weight to each edge of the graph. Graphs with weights, or weighted graphs, are used to represent structures in which pairwise connections have some numerical values. For example, if a graph represents a road network, the weights could represent the length of each road.

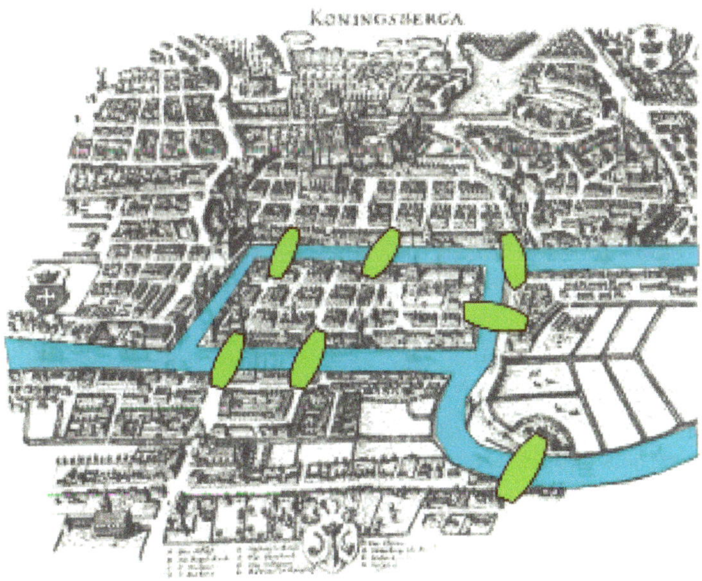

The Königsberg Bridge problem

The paper written by Leonhard Euler on the *Seven Bridges of Königsberg* and published in 1736 is regarded as the first paper in the history of graph theory. This paper, as well as the one written by Vandermonde on the *knight problem,* carried on with the *analysis situs* initiated by Leibniz. Euler›s formula relating the number of edges, vertices, and faces of a convex polyhedron was studied and generalized by Cauchy and L›Huillier, and represents the beginning of the branch of mathematics known as topology.

More than one century after Euler›s paper on the bridges of Königsberg and while Listing was introducing the concept of topology, Cayley was led by an interest in particular analytical forms arising from differential calculus to study a particular class of graphs, the *trees*. This study had many implications for theoretical chemistry. The techniques he used mainly concern the enumeration of graphs with particular properties. Enumerative graph theory then arose from the results of Cayley and the fundamental results published by Pólya between 1935 and 1937. These were generalized by De Bruijn in 1959. Cayley linked his results on trees with contemporary studies of chemical composition. The fusion of ideas from mathematics with those from chemistry began what has become part of the standard terminology of graph theory.

In particular, the term «graph» was introduced by Sylvester in a paper published in 1878 in *Nature*, where he draws an analogy between "quantic invariants" and "co-variants" of algebra and molecular diagrams:

"[...] Every invariant and co-variant thus becomes expressible by a *graph* precisely identical with a Kekuléan diagram or chemicograph. [...] I give a rule for the geometrical multiplication of graphs, *i.e.* for constructing a *graph* to the product of in- or co-variants whose separate graphs are given. [...]" (italics as in the original).

The first textbook on graph theory was written by Dénes Kőnig, and published in 1936. Another book by Frank Harary, published in 1969, was «considered the world over to be the definitive textbook on the subject», and enabled mathematicians, chemists, electrical engineers and social scientists to talk to each other. Harary donated all of the royalties to fund the Pólya Prize.

One of the most famous and stimulating problems in graph theory is the four color problem: «Is it true that any map drawn in the plane may have its regions colored with four colors, in such a way that any two regions having a common border have different colors?» This problem was first posed by Francis Guthrie in 1852 and its first written record is in a letter of De Morgan addressed to Hamilton the same year. Many incorrect proofs have been proposed, including those by Cayley, Kempe, and others. The study and the generalization of this problem by Tait, Heawood, Ramsey and Hadwiger led to the study of the colorings of the graphs embedded on surfaces with arbitrary genus. Tait›s reformulation generated a new class of problems, the *factorization problems*, particularly studied by Petersen and Kőnig. The works of Ramsey on colorations and more specially the results obtained by Turán in 1941 was at the origin of another branch of graph theory, *extremal graph theory*.

The four color problem remained unsolved for more than a century. In 1969 Heinrich Heesch published a method for solving the problem using computers. A computer-aided proof produced in 1976 by Kenneth Appel and Wolfgang Haken makes fundamental use of the notion of "discharging" developed by Heesch. The proof involved checking the properties of 1,936 configurations by computer, and was not fully accepted at the time due to its complexity. A simpler proof considering only 633 configurations was given twenty years later by Robertson, Seymour, Sanders and Thomas.

The autonomous development of topology from 1860 and 1930 fertilized graph theory back through the works of Jordan, Kuratowski and Whitney. Another important factor of common development of graph theory and topology came from the use of the techniques of modern algebra. The first example of such a use comes from the work of the physicist Gustav Kirchhoff, who published in 1845 his Kirchhoff›s circuit laws for calculating the voltage and current in electric circuits.

The introduction of probabilistic methods in graph theory, especially in the study of Erdős and Rényi of the asymptotic probability of graph connectivity, gave rise to yet another branch, known as *random graph theory*, which has been a fruitful source of graph-theoretic results.

GRAPH DRAWING

Graphs are represented visually by drawing a dot or circle for every vertex, and drawing an arc between two vertices if they are connected by an edge. If the graph is directed, the direction is indicated by drawing an arrow.

A graph drawing should not be confused with the graph itself (the abstract, non-visual structure) as there are several ways to structure the graph drawing. All that matters is which vertices are connected to which others by how many edges and not the exact layout. In practice it is often difficult to decide if two drawings represent the same graph. Depending on the problem domain some layouts may be better suited and easier to understand than others.

The pioneering work of W. T. Tutte was very influential in the subject of graph drawing. Among other achievements, he introduced the use of linear algebraic methods to obtain graph drawings.

Graph drawing also can be said to encompass problems that deal with the crossing number and its various generalizations. The crossing number of a graph is the minimum number of intersections between edges that a drawing of the graph in the plane must contain. For a planar graph, the crossing number is zero by definition.

Drawings on surfaces other than the plane are also studied.

GRAPH-THEORETIC DATA STRUCTURES

There are different ways to store graphs in a computer system. The data structure used depends on both the graph structure and the algorithm used for manipulating the graph. Theoretically one can distinguish between list and matrix structures but in concrete applications the best structure is often a combination of both. List structures are often preferred for sparse graphs as they have smaller memory requirements. Matrix structures on the other hand provide faster access for some applications but can consume huge amounts of memory.

List structures include the incidence list, an array of pairs of vertices, and the adjacency list, which separately lists the neighbors of each vertex: Much like the incidence list, each vertex has a list of which vertices it is adjacent to.

Matrix structures include the incidence matrix, a matrix of 0›s and 1›s whose rows represent vertices and whose columns represent edges, and the adjacency matrix, in which both the rows and columns are indexed by vertices. In both cases a 1 indicates two adjacent objects and a 0 indicates two non-adjacent objects. The Laplacian matrix is a modified form of the adjacency matrix that incorporates information about the degrees of the vertices, and is useful in some calculations such as Kirchhoff›s theorem on the number of spanning trees of a graph. The distance matrix, like the adjacency matrix, has both its rows and columns indexed by vertices, but rather than containing a 0 or a 1 in each cell it contains the length of a shortest path between two vertices.

PROBLEMS IN GRAPH THEORY

Enumeration

There is a large literature on graphical enumeration: the problem of counting graphs meeting specified conditions. Some of this work is found in Harary and Palmer (1973).

Subgraphs, induced Subgraphs, and Minors

A common problem, called the subgraph isomorphism problem, is finding a fixed graph as a subgraph in a given graph. One reason to be interested in such a question is that many graph properties are *hereditary* for subgraphs, which means that a graph has the property if and only if all subgraphs have it too. Unfortunately, finding maximal subgraphs of a certain kind is often an NP-complete problem. For example:

- Finding the largest complete subgraph is called the clique problem (NP-complete).

A similar problem is finding induced subgraphs in a given graph. Again, some important graph properties are hereditary with respect to induced subgraphs, which means that a graph has a property if and only if all induced subgraphs also have it. Finding maximal induced subgraphs of a certain kind is also often NP-complete. For example:

- Finding the largest edgeless induced subgraph or independent set is called the independent set problem (NP-complete).

Still another such problem, the minor containment problem, is to find a fixed graph as a minor of a given graph. A minor or subcontraction of a graph is any graph obtained by taking a subgraph and contracting some (or no) edges. Many graph properties are hereditary for minors, which means that a graph has a property if and only if all minors have it too. For example, Wagner›s Theorem states:

- A graph is planar if it contains as a minor neither the complete bipartite graph $K_{3,3}$ nor the complete graph K_5.

A similar problem, the subdivision containment problem, is to find a fixed graph as a subdivision of a given graph. A subdivision or homeomorphism of a graph is any graph obtained by subdividing some (or no) edges. Subdivision containment is related to graph properties such as planarity. For example, Kuratowski›s Theorem states:

- A graph is planar if it contains as a subdivision neither the complete bipartite graph $K_{3,3}$ nor the complete graph K_5.

Another problem in subdivision containment is Kelmans-Seymour conjecture:

- Every 5-vertex-connected graph that is not planar contains a subdivision of the 5-vertex complete graph K_5.

Another class of problems has to do with the extent to which various species and generalizations of graphs are determined by their *point-deleted subgraphs*. For example:

- The reconstruction conjecture

Graph Coloring

Many problems have to do with various ways of coloring graphs, for example:

- Four-color theorem
- Strong perfect graph theorem
- Erdős–Faber–Lovász conjecture (unsolved)
- Total coloring conjecture, also called Behzad›s conjecture (unsolved)
- List coloring conjecture (unsolved)
- Hadwiger conjecture (graph theory) (unsolved)

Subsumption and Unification

Constraint modeling theories concern families of directed graphs related by a partial order. In these applications, graphs are ordered by specificity, meaning that more constrained graphs—which are more specific and thus contain a greater amount of information—are subsumed by those that are more general. Operations between graphs include evaluating the direction of a subsumption relationship between two graphs, if any, and computing graph unification. The unification of two argument graphs is defined as the most general graph (or the computation thereof) that is consistent with (i.e. contains all of the information in) the inputs, if such a graph exists; efficient unification algorithms are known.

For constraint frameworks which are strictly compositional, graph unification is the sufficient satisfiability and combination function. Well-known applications include automatic theorem proving and modeling the elaboration of linguistic structure.

Route Problems

- Hamiltonian path problem
- Minimum spanning tree
- Route inspection problem (also called the "Chinese postman problem")
- Seven bridges of Königsberg
- Shortest path problem
- Steiner tree
- Three-cottage problem
- Traveling salesman problem (NP-hard)

Network Flow

There are numerous problems arising especially from applications that have to do with various notions of flows in networks, for example:

- Max flow min cut theorem

Visibility Problems

- Museum guard problem

Covering Problems

Covering problems in graphs are specific instances of subgraph-finding problems, and they tend to be closely related to the clique problem or the independent set problem.

- Set cover problem
- Vertex cover problem

Decomposition Problems

Decomposition, defined as partitioning the edge set of a graph (with as many vertices as necessary accompanying the edges of each part of the partition), has a wide variety of question. Often, it is required to decompose a graph into subgraphs isomorphic to a fixed graph; for instance, decomposing a complete graph into Hamiltonian cycles. Other problems specify a family of graphs into which a given graph should be decomposed, for instance, a family of cycles, or decomposing a complete graph K_n into $n - 1$ specified trees having, respectively, 1, 2, 3, ..., $n - 1$ edges.

Some specific decomposition problems that have been studied include:

- Arboricity, a decomposition into as few forests as possible
- Cycle double cover, a decomposition into a collection of cycles covering each edge exactly twice
- Edge coloring, a decomposition into as few matchings as possible
- Graph factorization, a decomposition of a regular graph into regular subgraphs of given degrees

Graph Classes

Many problems involve characterizing the members of various classes of graphs. Some examples of such questions are below:

- Enumerating the members of a class
- Characterizing a class in terms of forbidden substructures
- Ascertaining relationships among classes (e.g. does one property of graphs imply another)

- Finding efficient algorithms to decide membership in a class
- Finding representations for members of a class

REFERENCES

1. See, for instance, Iyanaga and Kawada, 69 J, p. 234 or Biggs, p. 4.
2. See, for instance, Graham et al., p. 5.
3. Hale, Scott A. (2013). "Multilinguals and Wikipedia Editing". arXiv:1312.0976Freely accessible [cs.CY].
4. Mashaghi, A.; et al. (2004). "Investigation of a protein complex network". European Physical Journal B. 41 (1): 113–121. doi:10.1140/epjb/e2004-00301-0.
5. Grandjean, Martin (2016). "A social network analysis of Twitter: Mapping the digital humanities community". Cogent Arts & Humanities. 3 (1): 1171458. doi:10.1080/23311983.2016.1171458.
6. Grandjean, Martin (2015). "Social network analysis and visualization: Moreno's Sociograms revisited". Redesigned network strictly based on Moreno (1934), Who Shall Survive.
7. Rosen, Kenneth H. Discrete mathematics and its applications (7th ed.). New York: McGraw-Hill. ISBN 978-0-07-338309-5.
8. Biggs, N.; Lloyd, E.; Wilson, R. (1986), Graph Theory, 1736-1936, Oxford University Press
9. Cauchy, A.L. (1813), "Recherche sur les polyèdres - premier mémoire", Journal de l'École Polytechnique, 9 (Cahier 16): 66–86.
10. L'Huillier, S.-A.-J. (1861), "Mémoire sur la polyèdrométrie", Annales de Mathématiques, 3: 169–189.
11. Cayley, A. (1857), "On the theory of the analytical forms called trees", Philosophical Magazine, Series IV, 13 (85): 172–176, doi:10.1017/CBO9780511703690.046
12. Cayley, A. (1875), "Ueber die Analytischen Figuren, welche in der Mathematik Bäume genannt werden und ihre Anwendung auf die Theorie chemischer Verbindungen", Berichte der deutschen Chemischen Gesellschaft, 8 (2): 1056–1059, doi:10.1002/cber.18750080252.
13. Sylvester, James Joseph (1878). "Chemistry and Algebra". Nature. 17: 284. doi:10.1038/017284a0.
14. Tutte, W.T. (2001), Graph Theory, Cambridge University Press, p. 30, ISBN 978-0-521-79489-3, retrieved 2016-03-14
15. Gardner, Martin (1992), Fractal Music, Hypercards, and more…Mathematical Recreations from Scientific American, W. H. Freeman and Company, p. 203
16. Society for Industrial and Applied Mathematics (2002), "The George Polya Prize", Looking Back, Looking Ahead: A SIAM History (PDF), p. 26, retrieved 2016-03-14

17. Heinrich Heesch: Untersuchungen zum Vierfarbenproblem. Mannheim: Bibliographisches Institut 1969.
18. Appel, K.; Haken, W. (1977), "Every planar map is four colorable. Part I. Discharging", Illinois J. Math., 21: 429–490.
19. Appel, K.; Haken, W. (1977), "Every planar map is four colorable. Part II. Reducibility", Illinois J. Math., 21: 491–567.
20. Robertson, N.; Sanders, D.; Seymour, P.; Thomas, R. (1997), "The four color theorem", Journal of Combinatorial Theory Series B, 70: 2–44, doi:10.1006/jctb.1997.1750
21. Berge, Claude (1958), Théorie des graphes et ses applications, Collection Universitaire de Mathématiques, II, Paris: Dunod. English edition, Wiley 1961; Methuen & Co, New York 1962; Russian, Moscow 1961; Spanish, Mexico 1962; Roumanian, Bucharest 1969; Chinese, Shanghai 1963; Second printing of the 1962 first English edition, Dover, New York 2001.
22. Biggs, N.; Lloyd, E.; Wilson, R. (1986), Graph Theory, 1736–1936, Oxford University Press.
23. Bondy, J.A.; Murty, U.S.R. (2008), Graph Theory, Springer, ISBN 978-1-84628-969-9.
24. Bollobás, Béla; Riordan, O.M (2003), Mathematical results on scale-free random graphs in "Handbook of Graphs and Networks" (S. Bornholdt and H.G. Schuster (eds)), Wiley VCH, Weinheim, 1st ed..
25. Chartrand, Gary (1985), Introductory Graph Theory, Dover, ISBN 0-486-24775-9.
26. Gibbons, Alan (1985), Algorithmic Graph Theory, Cambridge University Press.
27. Reuven Cohen, Shlomo Havlin (2010), Complex Networks: Structure, Robustness and Function, Cambridge University Press.
28. Golumbic, Martin (1980), Algorithmic Graph Theory and Perfect Graphs, Academic Press.
29. Harary, Frank (1969), Graph Theory, Reading, MA: Addison-Wesley.
30. Harary, Frank; Palmer, Edgar M. (1973), Graphical Enumeration, New York, NY: Academic Press.
31. Mahadev, N.V.R.; Peled, Uri N. (1995), Threshold Graphs and Related Topics, North-Holland.
32. Mark Newman (2010), Networks: An Introduction, Oxford University Press.

CHAPTER 5

Probability Theory and Distribution

Chapter Contents

- **Introduction**
- **History**
- **Treatment**
- **Classical Probability Distributions**
- **Convergence of Random Variables**
- **Probability Distribution**
- **Beta Distribution**

Probability theory is introduced in this unit. Experiments, outcomes, sample spaces, events, and conditional probability theory are covered.

INTRODUCTION

Probability theory is the branch of mathematics concerned with probability, the analysis of random phenomena. The central objects of probability theory are random variables, stochastic processes, and events: mathematical abstractions of non-deterministic events or measured quantities that may either be single occurrences or evolve over time in an apparently random fashion.

It is not possible to predict precisely results of random events. However, if a sequence of individual events, such as coin flipping or the roll of dice, is influenced by other factors, such as friction, it will exhibit certain patterns, which can be studied and predicted. Two representative mathematical results describing such patterns are the law of large numbers and the central limit theorem.

As a mathematical foundation for statistics, probability theory is essential to many human activities that involve quantitative analysis of large sets of data.Methods of probability theory also apply to descriptions of complex systems given only partial knowledge of their state, as in statistical mechanics. A great discovery of twentieth century physics was the probabilistic nature of physical phenomena at atomic scales, described in quantum mechanics.

HISTORY

The mathematical theory of probability has its roots in attempts to analyze games of chance by Gerolamo Cardano in the sixteenth century, and by Pierre de Fermat and Blaise Pascal in the seventeenth century (for example the "problem of points»). Christiaan Huygens published a book on the subject in 1657 and in the 19th century, Pierre Laplace completed what is today considered the classic interpretation.

Initially, probability theory mainly considered **discrete** events, and its methods were mainly combinatorial. Eventually, analytical considerations compelled the incorporation of **continuous** variables into the theory.

This culminated in modern probability theory, on foundations laid by Andrey Nikolaevich Kolmogorov. Kolmogorov combined the notion of sample space, introduced by Richard von Mises, and **measure theory** and presented his axiom system for probability theory in 1933. This became the mostly undisputed axiomatic basis for modern probability theory; but, alternatives exist, such as the adoption of finite rather than countable additivity by Bruno de Finetti.

TREATMENT

Most introductions to probability theory treat discrete probability distributions and continuous probability distributions separately. The more mathematically advanced measure theory-based treatment of probability covers the discrete, continuous, a mix of the two, and more.

Motivation

Consider an experiment that can produce a number of outcomes. The set of all outcomes is called the sample space of the experiment. The power set of the sample space (or equivalently, the event space) is formed by considering all different collections of possible results. For example, rolling an honest die produces one of six possible results. One collection of possible results corresponds to getting an odd number. Thus, the subset $\{1,3,5\}$ is an element of the power set of the sample space of die rolls. These collections are called events. In this case, $\{1,3,5\}$ is the event that the die falls on some odd number. If the results that actually occur fall in a given event, that event is said to have occurred.

Probability is a way of assigning every "event" a value between zero and one, with the requirement that the event made up of all possible results (in our example, the event $\{1,2,3,4,5,6\}$) be assigned a value of one. To qualify as a probability distribution, the assignment of values must satisfy the requirement that if you look at a collection of mutually exclusive events (events that contain no common results, e.g., the events {1,6}, {3}, and {2,4} are all mutually exclusive), the probability that one of the events will occur is given by the sum of the probabilities of the individual events.

The probability that any one of the events {1,6}, {3}, or {2,4} will occur is 5/6. This is the same as saying that the probability of event {1,2,3,4,6} is 5/6. This event encompasses the possibility of any number except five being rolled. The mutually exclusive event {5} has a probability of 1/6, and the event {1,2,3,4,5,6} has a probability of 1, that is, absolute certainty.

Discrete Probability Distributions

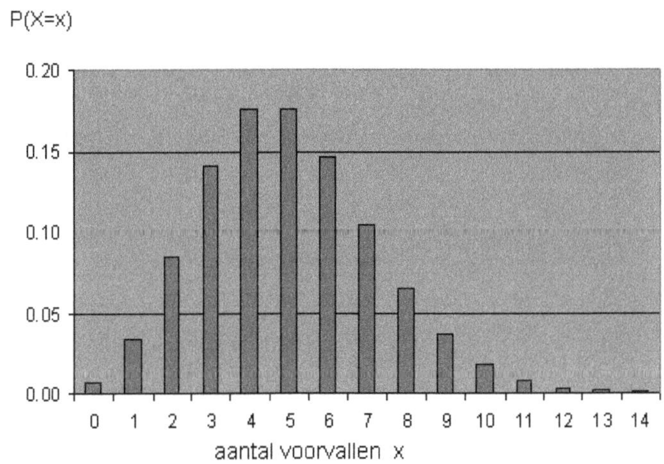

The Poisson distribution, a discrete probability distribution.

Discrete Probability Theory deals with events that occur in countable sample spaces.

Examples: Throwing dice, experiments with decks of cards, random walk, and tossing coins

Classical definition: Initially the probability of an event to occur was defined as number of cases favorable for the event, over the number of total outcomes possible in an equiprobable sample space: see Classical definition of probability.

For example, if the event is «occurrence of an even number when a die is rolled», the probability is given by $\frac{3}{6}=\frac{1}{2}$, since 3 faces out of the 6 have even numbers and each face has the same probability of appearing.

Modern definition: The modern definition starts with a finite or countable set called the **sample space**, which relates to the set of all possible outcomes in classical sense, denoted by Ω. It is then assumed that for each element $x \in \Omega$, an intrinsic "probability" value $f(x)$ is attached, which satisfies the following properties:

1. $f(x) \in [0,1]$ for all $x \in \Omega$;
2. $\sum_{x \in \Omega} f(x) = 1$.

That is, the probability function f(x) lies between zero and one for every value of x in the sample space Ω, and the sum of f(x) over all values x in the sample space Ω is equal to 1. An **event** is defined as any subset E of the sample space Ω. The **probability** of the event E is defined as

$$P(E) = \sum_{x \in E} f(x).$$

So, the probability of the entire sample space is 1, and the probability of the null event is 0.

The function $f(x)$ mapping a point in the sample space to the "probability" value is called a **probability mass function** abbreviated as **pmf**. The modern definition does not try to answer how probability mass functions are obtained; instead it builds a theory that assumes their existence.

Continuous Probability Distributions

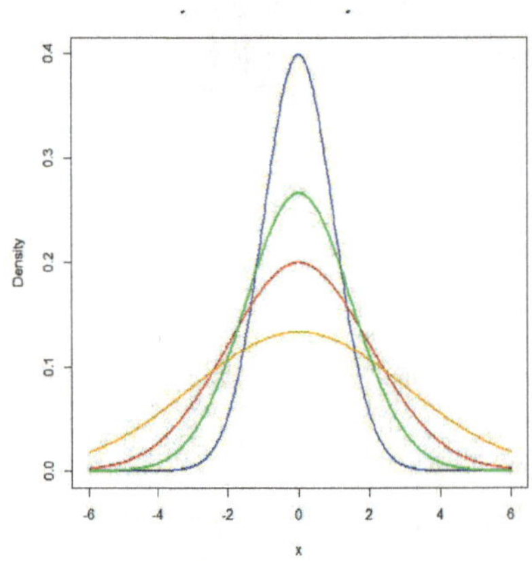

The normal distribution, a continuous probability distribution.

Continuous probability theory deals with events that occur in a continuous sample space.

Classical definition: The classical definition breaks down when confronted with the continuous case. See Bertrand›s paradox.

Modern definition: If the outcome space of a random variable X is the set of real numbers $(\mathbb{R})$ or a subset thereof, then a function called the **cumulative distribution**

function (or **cdf**) F exists, defined by $F(x) = P(X \leq x)$. That is, F(x) returns the probability that X will be less than or equal to x.

The cdf necessarily satisfies the following properties.

- F is a monotonically non-decreasing, right-continuous function;
- $\lim_{x\to-\infty} F(x) = 0$;
- $\lim_{x\to\infty} F(x) = 1$.

If F is absolutely continuous, i.e., its derivative exists and integrating the derivative gives us the cdf back again, then the random variable X is said to have a **probability density function** or **pdf** or simply **density** $f(x) = \frac{dF(x)}{dx}$

For a set $E \subseteq \mathbb{R}$, the probability of the random variable X being in E is

$$P(X \in E) = \int_{x\in E} dF(x).$$

In case the probability density function exists, this can be written as

$$P(X \in E) = \int_{x\in E} f(x)\, dx.$$

Whereas the pdf exists only for continuous random variables, the cdf exists for all random variables (including discrete random variables) that take values in $\mathbb{R}$

These concepts can be generalized for multidimensional cases on $\mathbb{R}^n$ and other continuous sample spaces.

Measure-Theoretic Probability Theory

The raison d'être of the measure-theoretic treatment of probability is that it unifies the discrete and the continuous cases, and makes the difference a question of which measure is used. Furthermore, it covers distributions that are neither discrete nor continuous nor mixtures of the two.

An example of such distributions could be a mix of discrete and continuous distributions—for example, a random variable that is 0 with probability 1/2, and takes a random value from a normal distribution with probability 1/2. It can still be studied to some extent by considering it to have a pdf of is the Dirac delta function.

Other distributions may not even be a mix, for example, the Cantor distribution has no positive probability for any single point, neither does it have a density. The modern approach to probability theory solves these problems using measure theory to define the probability space:

Given any set Ω, (also called **sample space**) and a σ-algebra F on it, a measure P on F is called a **probability measure** if $P(\Omega) = 1$.

If F is the Borel σ-algebra on the set of real numbers, then there is a unique probability measure on $\mathcal{F}$ for any cdf, and vice versa. The measure corresponding to a cdf is said to be **induced** by the cdf. This measure coincides with the pmf for discrete variables and pdf for continuous variables, making the measure-theoretic approach free of fallacies.

The probability of a set E in the σ-algebra $\mathcal{F}$ is defined as

$$P(E) = \int_{\omega \in E} \mu_F(d\omega)$$

where the integration is with respect to the measure μ_F induced by F

Along with providing better understanding and unification of discrete and continuous probabilities, measure-theoretic treatment also allows us to work on probabilities outside $\mathbb{R}^n$, as in the theory of stochastic processes. For example, to study Brownian motion, probability is defined on a space of functions.

When it's convenient to work with a dominating measure, the Radon-Nikodym theorem is used to define a density as the Radon-Nikodym derivative of the probability distribution of interest with respect to this dominating measure. Discrete densities are usually defined as this derivative with respect to a counting measure over the set of all possible outcomes. Densities for absolutely continuous distributions are usually defined as this derivative with respect to the Lebesgue measure. If a theorem can be proved in this general setting, it holds for both discrete and continuous distributions as well as others; separate proofs are not required for discrete and continuous distributions.

CLASSICAL PROBABILITY DISTRIBUTIONS

Certain random variables occur very often in probability theory because they well describe many natural or physical processes. Their distributions therefore have gained special importance in probability theory. Some fundamental discrete distributions are the discrete uniform, Bernoulli, binomial, negative binomial, Poisson and geometric distributions. Important continuous distributions include the continuous uniform, normal, exponential, gamma and beta distributions.

CONVERGENCE OF RANDOM VARIABLES

In probability theory, there are several notions of convergence for random variables. They are listed below in the order of strength, i.e., any subsequent notion of convergence in the list implies convergence according to all of the preceding notions.

Weak Convergence

A sequence of random variables $X_1, X_2, \dots,$ converges **weakly** to the random variable X if their respective cumulative distribution functions $F_1, F_2, \dots$ converge to the cumulative distribution function F of X, wherever F is continuous. Weak convergence is also called **convergence in distribution**.

Most common shorthand notation: $X_n \xrightarrow{\mathcal{D}} X$

Convergence in Probability

The sequence of random variables $X_1, X_2, \cdots$ is said to converge towards the random variable X **in probability** if $\lim_{n\to\infty} P(|X_n - X| \geq \varepsilon) = 0$ for every ε > 0.

Most common shorthand notation: $X_n \xrightarrow{P} X$

Strong convergence

The sequence of random variables $X_1, X_2, \ldots$ is said to converge towards the random variable X **strongly** if $P(\lim_{n\to\infty} X_n = X) = 1$. Strong convergence is also known as **almost sure convergence**.

Most common shorthand notation: $X_n \xrightarrow{\text{a.s.}} X$

As the names indicate, weak convergence is weaker than strong convergence. In fact, strong convergence implies convergence in probability, and convergence in probability implies weak convergence. The reverse statements are not always true.

Law of Large Numbers

Common intuition suggests that if a fair coin is tossed many times, then roughly half of the time it will turn up heads, and the other half it will turn up tails. Furthermore, the more often the coin is tossed, the more likely it should be that the ratio of the number of heads to the number of tails will approach unity. Modern probability theory provides a formal version of this intuitive idea, known as the **law of large numbers**. This law is remarkable because it is not assumed in the foundations of probability theory, but instead emerges from these foundations as a theorem. Since it links theoretically derived probabilities to their actual frequency of occurrence in the real world, the law of large numbers is considered as a pillar in the history of statistical theory and has had widespread influence.

The **law of large numbers** (LLN) states that the sample average

$$\overline{X}_n = \frac{1}{n}\sum_{k=1}^{n} X_k$$

of a sequence of independent and identically distributed random variables X_k converges towards their common expectation μ, provided that the expectation of $|X_k|$ is finite.

It is in the different forms of convergence of random variables that separates the weak and the strong law of large numbers

Weak law: $\overline{X}_n \xrightarrow{P} \mu$ for $n \to \infty$

Strong law: $\overline{X}_n \xrightarrow{\text{a.s.}} \mu$ for $n \to \infty$.

It follows from the LLN that if an event of probability p is observed repeatedly during independent experiments, the ratio of the observed frequency of that event to the total number of repetitions converges towards p.

For example, if $Y_1, Y_2, \ldots$ are independent Bernoulli random variables taking values 1 with probability p and 0 with probability 1-p, then $\mathrm{E}(Y_i) = p$ for all i, so that $\bar{Y}_n$ converges to p almost surely.

Central Limit Theorem

"The central limit theorem (CLT) is one of the great results of mathematics." (Chapter 18 in) It explains the ubiquitous occurrence of the normal distribution in nature.

The theorem states that the average of many independent and identically distributed random variables with finite variance tends towards a normal distribution irrespective of the distribution followed by the original random variables. Formally, let $X_1, X_2, \ldots$ be independent random variables with mean μ and variance $\sigma^2 > 0$ Then the sequence of random variables

$$Z_n = \frac{\sum_{i=1}^{n}(X_i - \mu)}{\sigma\sqrt{n}}$$

converges in distribution to a standard normal random variable.

For some classes of random variables the classic central limit theorem works rather fast (see Berry–Esseen theorem), for example the distributions with finite first, second and third moment from the exponential family; on the other hand, for some random variables of the heavy tail and fat tail variety, it works very slowly or may not work at all: in such cases one may use the Generalized Central Limit Theorem (GCLT).

PROBABILITY DISTRIBUTION

In probability and statistics, a **probability distribution** is a mathematical function that, stated in simple terms, can be thought of as providing the probability of occurrence of different possible outcomes in an experiment. For instance, if the random variable X is used to denote the outcome of a coin toss ('the experiment'), then the probability distribution of X would take the value 0.5 for $X = \text{Heads}$, and 0.5 for $X = \text{Tails}$.

In more technical terms, the probability distribution is a description of a random phenomenon in terms of the probabilities of events. Examples of random phenomena can include the results of an experiment or survey. A probability distribution is defined in terms of an underlying sample space, which is the set of all possible outcomes of the random phenomenon being observed. The sample space may be the set of real numbers or a higher-dimensional vector space, or it may be a list of non-numerical values; for example, the sample space of a coin flip would be $\{\text{Heads}, \text{Tails}\}$.

Probability distributions are generally divided into two classes. A **discrete probability distribution** (applicable to the scenario where the set of possible outcomes is discrete, such as a coin toss or a roll of dice) can be encoded by a discrete list of the probabilities of the outcomes, known as a probability mass function. On the other hand, a **continuous probability distribution** (applicable to the scenarios where the set of possible outcomes can take on values in a continuous range (e.g., real numbers), such as the temperature on a given day) is typically described by probability density functions (with the probability of any individual outcome actually being 0). The normal distribution represents a commonly encountered continuous probability distribution. More complex experiments, such as those involving stochastic processes defined in continuous time, may demand the use of more general probability measures.

A probability distribution whose sample space is the set of real numbers is called univariate, while a distribution whose sample space is a vector space is called multivariate. A univariate distribution gives the probabilities of a single random variable taking on various alternative values; a multivariate distribution (a joint probability distribution) gives the probabilities of a random vector—a list of two or more random variables—taking on various combinations of values. Important and commonly encountered univariate probability distributions include the binomial distribution, the hypergeometric distribution, and the normal distribution. The multivariate normal distribution is a commonly encountered multivariate distribution.

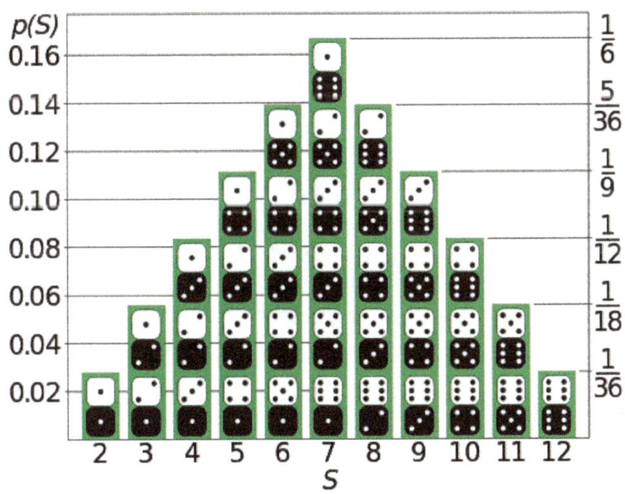

The probability mass function (pmf) p(S) specifies the probability distribution for the sum S of counts from two dice. For example, the figure shows that p(11) = 1/18. The pmf allows the computation of probabilities of events such as P(S > 9) = 1/12 + 1/18 + 1/36 = 1/6, and all other probabilities in the distribution.

To define probability distributions for the simplest cases, one needs to distinguish between **discrete** and **continuous** random variables. In the discrete case, it is sufficient to specify a probability mass function assigning a probability to each possible outcome: for example, when throwing a fair dice, each of the six values 1 to 6 has the probability

1/6. The probability of an event is then defined to be the sum of the probabilities of the outcomes that satisfy the event; for example, the probability of the event "the die rolls an even value" is

$$\text{Prob}(2) + \text{Prob}(4) + \text{Prob}(6) = 1/6 + 1/6 + 1/6 = 1/2.$$

In contrast, when a random variable takes values from a continuum then typically, any individual outcome has probability zero and only events that include infinitely many outcomes, such as intervals, can have positive probability. For example, the probability that a given object weighs exactly 500 g is zero, because the probability of measuring exactly 500 g tends to zero as the accuracy of our measuring instruments increases. Nevertheless, in quality control one might demand that the probability of a "500 g" package containing between 490 g and 510 g should be no less than 98%, and this demand is less sensitive to the accuracy of our instruments.

Continuous probability distributions can be described in several ways. The probability density function describes the infinitesimal probability of any given value, and the probability that the outcome lies in a given interval can be computed by integrating the probability density function over that interval. On the other hand, the cumulative distribution function describes the probability that the random variable is no larger than a given value; the probability that the outcome lies in a given interval can be computed by taking the difference between the values of the cumulative distribution function at the endpoints of the interval. The cumulative distribution function is the antiderivative of the probability density function provided that the latter function exists.

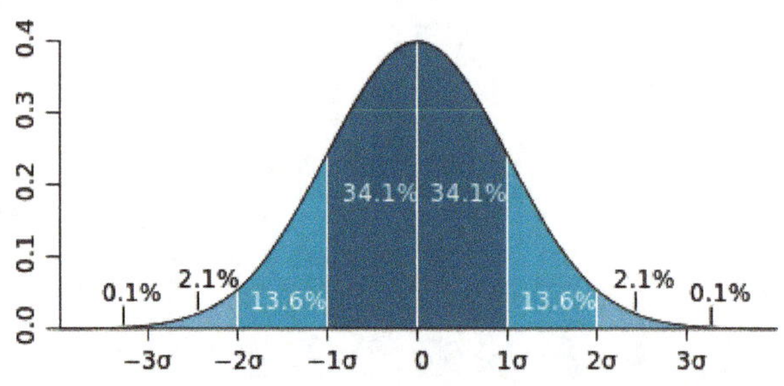

The probability density function (pdf) of the normal distribution, also called Gaussian or «bell curve», the most important continuous random distribution. As notated on the figure, the probabilities of intervals of values correspond to the area under the curve.

Terminology

As probability theory is used in quite diverse applications, terminology is not uniform and sometimes confusing. The following terms are used for non-cumulative probability distribution functions:

- **Probability mass**, Probability mass function, **p.m.f.**: for discrete random variables.
- Categorical distribution: for discrete random variables with a finite set of values.

- **Probability density**, Probability density function, **p.d.f.**: most often reserved for continuous random variables.

The following terms are somewhat ambiguous as they can refer to non-cumulative or cumulative distributions, depending on authors' preferences:

- **Probability distribution function**: continuous or discrete, non-cumulative or cumulative.
- **Probability function**: even more ambiguous, can mean any of the above or other things.

Finally,

- **Probability distribution**: sometimes the same as probability distribution function, but usually refers to the more complete assignment of probabilities to all measurable subsets of outcomes (i.e. the corresponding probability measure), not just to specific outcomes or ranges of outcomes.

Basic Terms

- Mode: for a discrete random variable, the value with highest probability (the location at which the probability mass function has its peak); for a continuous random variable, the location at which the probability density function has its peak.
- Support: the smallest closed set whose complement has probability zero.
- Head: the range of values where the pmf or pdf is relatively high.
- Tail: the complement of the head within the support; the large set of values where the pmf or pdf is relatively low.
- Expected value or **mean**: the weighted average of the possible values, using their probabilities as their weights; or the continuous analog thereof.
- Median: the value such that the set of values less than the median has a probability of one-half.
- Variance: the second moment of the pmf or pdf about the mean; an important measure of the dispersion of the distribution.
- Standard deviation: the square root of the variance, and hence another measure of dispersion.
- **Symmetry**: a property of some distributions in which the portion of the distribution to the left of a specific value is a mirror image of the portion to its right.
- Skewness: a measure of the extent to which a pmf or pdf "leans" to one side of its mean. The third standardized moment of the distribution.
- Kurtosis: a measure of the «fatness» of the tails of a pmf or pdf. The fourth standardized moment of the distribution.

Cumulative Distribution Function

Because a probability distribution Pr on the real line is determined by the probability of a scalar random variable X being in a half-open interval (-∞, x], the probability distribution is completely characterized by its cumulative distribution function:

$$F(x) = \Pr[X \le x] \qquad \text{for all } x \in \mathbb{R}.$$

Discrete Probability Distribution

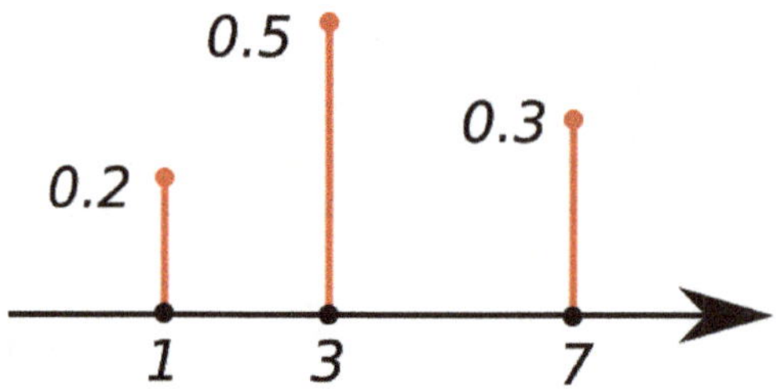

The probability mass function of a discrete probability distribution. The probabilities of the singletons {1}, {3}, and {7} are respectively 0.2, 0.5, 0.3. A set not containing any of these points has probability zero.

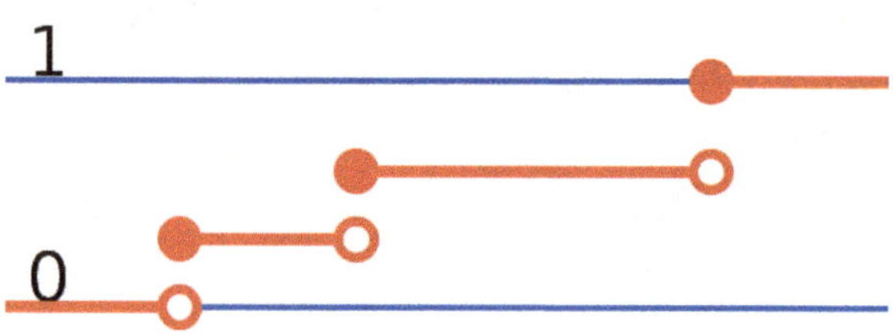

The cdf of a discrete probability distribution, ...

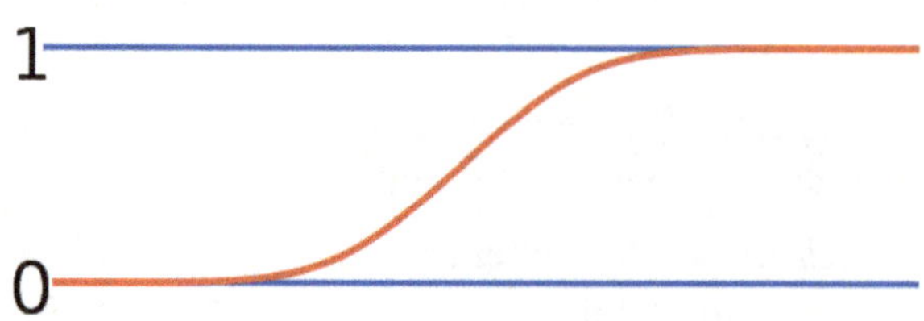

... of a continuous probability distribution, ...

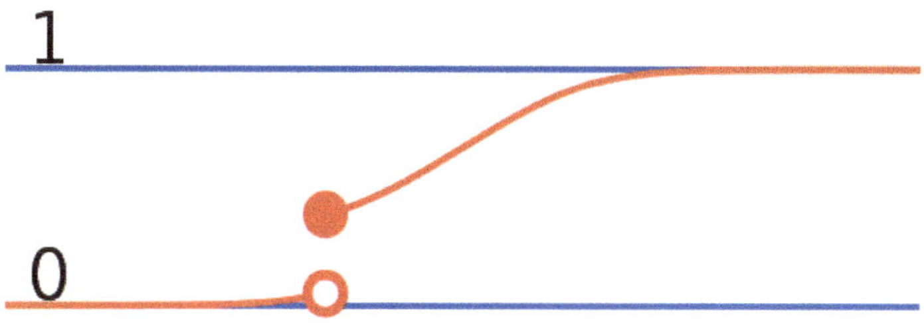

... of a distribution which has both a continuous part and a discrete part.

A **discrete probability distribution** is a probability distribution characterized by a probability mass function. Thus, the distribution of a random variable X is discrete, and X is called a **discrete random variable**, if

$$\sum_u \Pr(X = u) = 1$$

as u runs through the set of all possible values of X. A discrete random variable can assume only a finite or countably infinite number of values. For the number of potential values to be countably infinite, even though their probabilities sum to 1, the probabilities have to decline to zero fast enough. For example, if $\Pr(X = n) = \frac{1}{2^n}$ for $n = 1, 2, \ldots,$ we have the sum of probabilities 1/2 + 1/4 + 1/8 + ... = 1.

Well-known discrete probability distributions used in statistical modeling include the Poisson distribution, the Bernoulli distribution, the binomial distribution, the geometric distribution, and the negative binomial distribution. Additionally, the discrete uniform distribution is commonly used in computer programs that make equal-probability random selections between a number of choices.

Measure Theoretic Formulation

A measurable function $X\colon A \to B$ between a probability space $(A, \mathcal{A}, P)$ and a measurable space $(B, \mathcal{B})$ is called a **discrete random variable** provided its image is a countable set and the pre-image of singleton sets are measurable, i.e., $X^{-1}(b) \in \mathcal{A}$ for all $b \in B$. The latter requirement induces a probability mass function $f_X\colon X(A) \to \mathbb{R}$ via $f_X(b) := P(X^{-1}(b))$. Since the pre-images of disjoint sets are disjoint

$$\sum_{b \in X(A)} f_X(b) = \sum_{b \in X(A)} P(X^{-1}(b)) = P\left(\bigcup_{b \in X(A)} X^{-1}(b)\right) = P(A) = 1.$$

This recovers the definition given above.

Cumulative Density

Equivalently to the above, a discrete random variable can be defined as a random variable whose cumulative distribution function (cdf) increases only by jump discontinuities—that is, its cdf increases only where it «jumps» to a higher value, and is constant between those jumps. The points where jumps occur are precisely the values which the random variable may take.

Delta-function representation

Consequently, a discrete probability distribution is often represented as a generalized probability density function involving Dirac delta functions, which substantially unifies the treatment of continuous and discrete distributions. This is especially useful when dealing with probability distributions involving both a continuous and a discrete part.

Indicator-function representation

For a discrete random variable X, let u_0, u_1, ... be the values it can take with non-zero probability. Denote

$\Omega_i = X^{-1}(u_i) = \{\omega : X(\omega) = u_i\},\ i = 0, 1, 2, \ldots$

These are disjoint sets, and by formula (1)

$$\Omega_i = X^{-1}(u_i) = \{\omega : X(\omega) = u_i\},\ i = 0, 1, 2, \ldots$$

It follows that the probability that X takes any value except for u_0, u_1, ... is zero, and thus one can write X as

$$X(\omega) = \sum_i u_i 1_{\Omega_i}(\omega)$$

except on a set of probability zero, where A is the indicator function of A. This may serve as an alternative definition of discrete random variables.

Continuous Probability Distribution

A **continuous probability distribution** is a probability distribution that has a cumulative distribution function that is continuous. Most often they are generated by having a probability density function. Mathematicians call distributions with probability density functions **absolutely continuous**, since their cumulative distribution function is absolutely continuous with respect to the Lebesgue measure λ. If the distribution of X is continuous, then X is called a **continuous random variable**. There are many examples of continuous probability distributions: normal, uniform, chi-squared, and others.

Intuitively, a continuous random variable is the one which can take a continuous range of values—as opposed to a discrete distribution, where the set of possible values for the random variable is at most countable. While for a discrete distribution an event with probability zero is impossible (e.g., rolling 31/2 on a standard dice is impossible, and has probability zero), this is not so in the case of a continuous random variable. For example, if one measures the width of an oak leaf, the result of 3½ cm is possible; however, it has probability zero because uncountably many other potential values exist even between 3 cm and 4 cm. Each of these individual outcomes has probability zero, yet the probability that the outcome will fall into the interval (3 cm, 4 cm) is nonzero.

This apparent paradox is resolved by the fact that the probability that X attains some value within an infinite set, such as an interval, cannot be found by naively adding the probabilities for individual values. Formally, each value has an infinitesimally small probability, which statistically is equivalent to zero.

Formally, if X is a continuous random variable, then it has a probability density function *f*(x), and therefore its probability of falling into a given interval, say [a, b] is given by the integral

$$\Pr[a \leq X \leq b] = \int_a^b f(x)\, dx$$

In particular, the probability for X to take any single value a (that is a ≤ X ≤ a) is zero, because an integral with coinciding upper and lower limits is always equal to zero.

The definition states that a continuous probability distribution must possess a density, or equivalently, its cumulative distribution function be absolutely continuous. This requirement is stronger than simple continuity of the cumulative distribution function, and there is a special class of distributions, singular distributions, which are neither continuous nor discrete nor a mixture of those. An example is given by the Cantor distribution. Such singular distributions however are never encountered in practice.

Note on terminology: some authors use the term «continuous distribution» to denote the distribution with continuous cumulative distribution function. Thus, their definition includes both the (absolutely) continuous and singular distributions.

By one convention, a probability distribution μ is called continuous if its cumulative distribution function $F(x) = \mu(-\infty, x]$ is continuous and, therefore, the probability measure of singletons $\mu\{x\} = 0$ for all x.

Another convention reserves the term continuous probability distribution for absolutely continuous distributions. These distributions can be characterized by a probability density function: a non-negative Lebesgue integrable function {\displaystyle \,f} defined on the real numbers such that

$$F(x) = \mu(-\infty, x] = \int_{-\infty}^{x} f(t)\, dt.$$

Discrete distributions and some continuous distributions (like the Cantor distribution) do not admit such a density.

Some Properties

- The probability distribution of the sum of two independent random variables is the **convolution** of each of their distributions.
- Probability distributions are not a vector space—they are not closed under linear combinations, as these do not preserve non-negativity or total integral 1—but they are closed under convex combination, thus forming a convex subset of the space of functions (or measures).

Kolmogorov Definition

In the measure-theoretic formalization of probability theory, a random variable is defined as a measurable function X from a probability space $(\Omega,\mathcal{F},P)$ to measurable space $(\mathcal{X},\mathcal{A})$. A **probability distribution of** ***X*** is the pushforward measure X_*P of X , which is a probability measure on $X_*P = PX^{-1}$ satisfying $X_*P = PX^{-1}$.

Random Number Generation

A frequent problem in statistical simulations (the Monte Carlo method) is the generation of pseudo-random numbers that are distributed in a given way. Most algorithms are based on a pseudorandom number generator that produces numbers X that are uniformly distributed in the interval [0,1]. These random variates X are then transformed via some algorithm to create a new random variate having the required probability distribution.

Applications

The concept of the probability distribution and the random variables which they describe underlies the mathematical discipline of probability theory, and the science of statistics. There is spread or variability in almost any value that can be measured in a population (e.g. height of people, durability of a metal, sales growth, traffic flow, etc.); almost all measurements are made with some intrinsic error; in physics many processes are described probabilistically, from the kinetic properties of gases to the quantum mechanical description of fundamental particles. For these and many other reasons, simple numbers are often inadequate for describing a quantity, while probability distributions are often more appropriate.

As a more specific example of an application, the cache language models and other statistical language models used in natural language processing to assign probabilities to the occurrence of particular words and word sequences do so by means of probability distributions.

Common Probability Distributions

The following is a list of some of the most common probability distributions, grouped by the type of process that they are related to. For a more complete list, see list of probability distributions, which groups by the nature of the outcome being considered (discrete, continuous, multivariate, etc.)

Note also that all of the univariate distributions below are singly peaked; that is, it is assumed that the values cluster around a single point. In practice, actually observed quantities may cluster around multiple values. Such quantities can be modeled using a mixture distribution.

Related to real-valued quantities that grow linearly (e.g. errors, offsets)

- Normal distribution (Gaussian distribution), for a single such quantity; the most common continuous distribution

Related to positive real-valued quantities that grow exponentially (e.g. prices, incomes, populations)

- Log-normal distribution, for a single such quantity whose log is normally distributed
- Pareto distribution, for a single such quantity whose log is exponentially distributed; the prototypical power law distribution

Related to Real-Valued Quantities that are assumed to be Uniformly Distributed over a (Possibly Unknown) Region

- Discrete uniform distribution, for a finite set of values (e.g. the outcome of a fair die)
- Continuous uniform distribution, for continuously distributed values

Related to Bernoulli trials (yes/no events, with a Given Probability)

- Basic distributions:
    - Bernoulli distribution, for the outcome of a single Bernoulli trial (e.g. success/failure, yes/no)
    - Binomial distribution, for the number of "positive occurrences" (e.g. successes, yes votes, etc.) given a fixed total number of independent occurrences
    - Negative binomial distribution, for binomial-type observations but where the quantity of interest is the number of failures before a given number of successes occurs
    - Geometric distribution, for binomial-type observations but where the quantity of interest is the number of failures before the first success; a special case of the negative binomial distribution
- Related to sampling schemes over a finite population:
    - Hypergeometric distribution, for the number of «positive occurrences» (e.g. successes, yes votes, etc.) given a fixed number of total occurrences, using sampling without replacement
    - Beta-binomial distribution, for the number of «positive occurrences» (e.g. successes, yes votes, etc.) given a fixed number of total occurrences, sampling using a Polya urn scheme (in some sense, the "opposite" of sampling without replacement)

Related to Categorical Outcomes (events with K Possible Outcomes, with a given Probability for each Outcome)

- Categorical distribution, for a single categorical outcome (e.g. yes/no/maybe in a survey); a generalization of the Bernoulli distribution
- Multinomial distribution, for the number of each type of categorical outcome, given a fixed number of total outcomes; a generalization of the binomial distribution
- Multivariate hypergeometric distribution, similar to the multinomial distribution, but using sampling without replacement; a generalization of the hypergeometric distribution

Related to events in a Poisson Process (events that occur independently with a given rate)

- Poisson distribution, for the number of occurrences of a Poisson-type event in a given period of time
- Exponential distribution, for the time before the next Poisson-type event occurs
- Gamma distribution, for the time before the next k Poisson-type events occur

Related to the Absolute Values of Vectors with Normally Distributed Components

- Rayleigh distribution, for the distribution of vector magnitudes with Gaussian distributed orthogonal components. Rayleigh distributions are found in RF signals with Gaussian real and imaginary components.
- Rice distribution, a generalization of the Rayleigh distributions for where there is a stationary background signal component. Found in Rician fading of radio signals due to multipath propagation and in MR images with noise corruption on non-zero NMR signals.

Related to Normally Distributed Quantities Operated with Sum of Squares (for Hypothesis Testing)

- Chi-squared distribution, the distribution of a sum of squared standard normal variables; useful e.g. for inference regarding the sample variance of normally distributed samples (see chi-squared test)
- Student›s t distribution, the distribution of the ratio of a standard normal variable and the square root of a scaled chi squared variable; useful for inference regarding the mean of normally distributed samples with unknown variance (see Student›s t-test)
- F-distribution, the distribution of the ratio of two scaled chi squared variables; useful e.g. for inferences that involve comparing variances or involving R-squared (the squared correlation coefficient)

Useful as Conjugate Prior Distributions in Bayesian Inference

- Beta distribution, for a single probability (real number between 0 and 1); conjugate to the Bernoulli distribution and binomial distribution
- Gamma distribution, for a non-negative scaling parameter; conjugate to the rate parameter of a Poisson distribution or exponential distribution, the precision (inverse variance) of a normal distribution, etc.
- Dirichlet distribution, for a vector of probabilities that must sum to 1; conjugate to the categorical distribution and multinomial distribution; generalization of the beta distribution
- Wishart distribution, for a symmetric non-negative definite matrix; conjugate to the inverse of the covariance matrix of a multivariate normal distribution; generalization of the gamma distribution

BETA DISTRIBUTION

In probability theory and statistics, the **beta distribution** is a family of continuous probability distributions defined on the interval [0, 1] parametrized by two positive shape parameters, denoted by α and β, that appear as exponents of the random variable and control the shape of the distribution.

The beta distribution has been applied to model the behavior of random variables limited to intervals of finite length in a wide variety of disciplines. For example, it has been used as a statistical description of allele frequencies in population genetics; time allocation in project management / control systems; sunshine data; variability of soil properties; proportions of the minerals in rocks in stratigraphy; and heterogeneity in the probability of HIV transmission.

In Bayesian inference, the beta distribution is the conjugate prior probability distribution for the Bernoulli, binomial, negative binomial and geometric distributions. For example, the beta distribution can be used in Bayesian analysis to describe initial knowledge concerning probability of success such as the probability that a space vehicle will successfully complete a specified mission. The beta distribution is a suitable model for the random behavior of percentages and proportions.

The usual formulation of the beta distribution is also known as the **beta distribution of the first kind**, whereas beta distribution of the second kind is an alternative name for the beta prime distribution.

Characterization

Probability Density Function

The probability density function (pdf) of the beta distribution, for $0 \leq x \leq 1$, and shape parameters α, $\beta > 0$, is a power function of the variable x and of its reflection $(1 - x)$ as follows:

$$\begin{aligned} f(x;\alpha,\beta) &= \text{constant}\cdot x^{\alpha-1}(1-x)^{\beta-1} \\ &= \frac{x^{\alpha-1}(1-x)^{\beta-1}}{\int_0^1 u^{\alpha-1}(1-u)^{\beta-1}\,du} \\ &= \frac{\Gamma(\alpha+\beta)}{\Gamma(\alpha)\Gamma(\beta)}\,x^{\alpha-1}(1-x)^{\beta-1} \\ &= \frac{1}{\mathrm{B}(\alpha,\beta)}\,x^{\alpha-1}(1-x)^{\beta-1} \end{aligned}$$

where Γ(z) is the gamma function. The beta function, B, is a normalization constant to ensure that the total probability integrates to 1. In the above equations x is a realization—an observed value that actually occurred—of a random process X.

This definition includes both ends x = 0 and x = 1, which is consistent with definitions for other continuous distributions supported on a bounded interval which are special cases of the beta distribution, for example the arcsine distribution, and consistent with several authors, like N. L. Johnson and S. Kotz. However, the inclusion of x = 0 and x = 1 does not work for $\alpha, \beta < 1$; accordingly, several other authors, including W. Feller, choose to exclude the ends x = 0 and x = 1, (such that the two ends are not actually part of the density function) and consider instead 0 < x < 1.

Several authors, including N. L. Johnson and S. Kotz, use the symbols p and q (instead of α and β) for the shape parameters of the beta distribution, reminiscent of the symbols traditionally used for the parameters of the Bernoulli distribution, because the beta distribution approaches the Bernoulli distribution in the limit when both shape parameters α and β approach the value of zero.

In the following, a random variable X beta-distributed with parameters α and β will be denoted by:

$$X \sim \mathrm{Beta}(\alpha,\beta)$$

Other notations for beta-distributed random variables used in the statistical literature are $X \sim \mathcal{B}e(\alpha,\beta)$ and $X \sim \beta_{\alpha,\beta}$.

Differential Equation

The probability density function satisfies the differential equation

$$f'(x) = f(x)\,\frac{(\alpha+\beta-2)x-(\alpha-1)}{(x-1)x}.$$

Cumulative Distribution Function

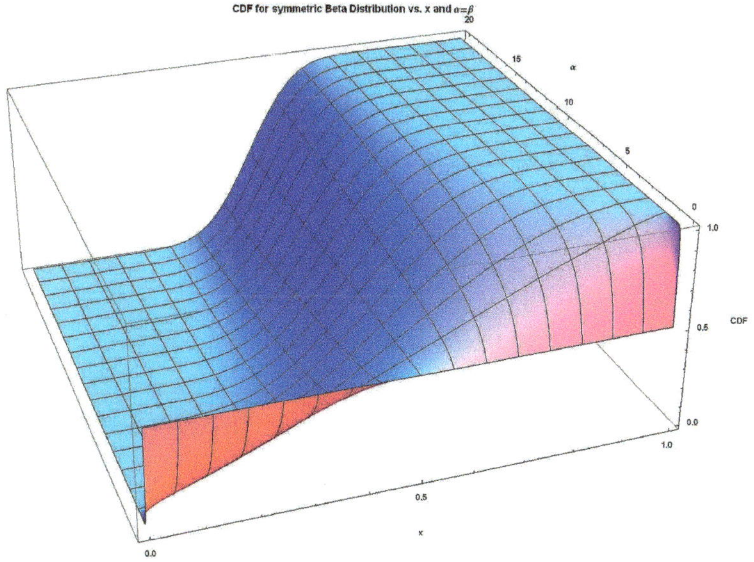

CDF for symmetric beta distribution vs. x and alpha=beta.

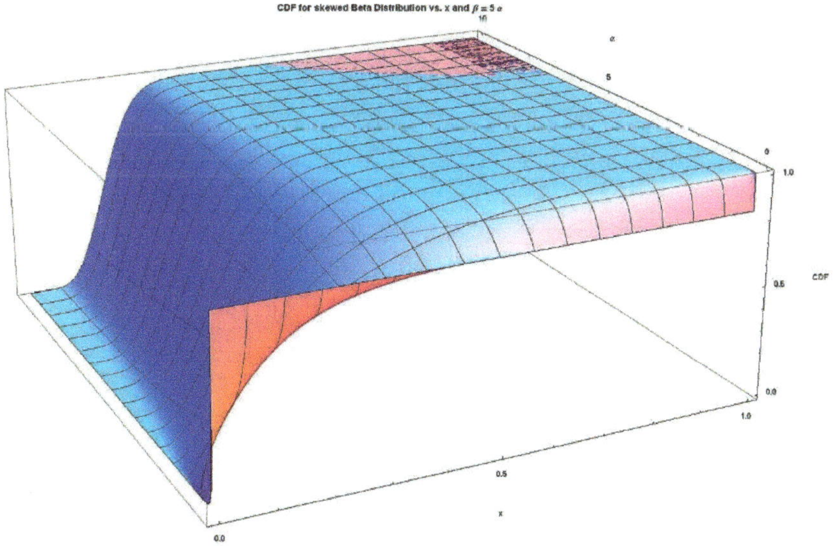

CDF for skewed beta distribution vs. x and beta= 5 alpha.

The cumulative distribution function is

$$F(x;\alpha,\beta) = \frac{\mathrm{B}(x;\alpha,\beta)}{\mathrm{B}(\alpha,\beta)} = I_x(\alpha,\beta)$$

where $B(x;\alpha,\beta)$ is the incomplete beta function and $I_x(\alpha,\beta)$ is the regularized incomplete beta function.

Properties

Measures of Central Tendency

Mode

The mode of a Beta distributed random variable X with α, β > 1 is the most likely value of the distribution (corresponding to the peak in the PDF), and is given by the following expression:

$$\frac{\alpha-1}{\alpha+\beta-2}.$$

When both parameters are less than one (α, β < 1), this is the anti-mode: the lowest point of the probability density curve.

Letting α = β, the expression for the mode simplifies to 1/2, showing that for α = β > 1 the mode (resp. anti-mode when α, β < 1), is at the center of the distribution: it is symmetric in those cases. See «Shapes» section in this article for a full list of mode cases, for arbitrary values of α and β. For several of these cases, the maximum value of the density function occurs at one or both ends. In some cases the (maximum) value of the density function occurring at the end is finite. For example, in the case of α = 2, β = 1 (or α = 1, β = 2), the density function becomes a right-triangle distribution which is finite at both ends. In several other cases there is a singularity at one end, where the value of the density function approaches infinity. For example, in the case α = β = 1/2, the Beta distribution simplifies to become the arcsine distribution. There is debate among mathematicians about some of these cases and whether the ends (x = 0, and x = 1) can be called modes or not.

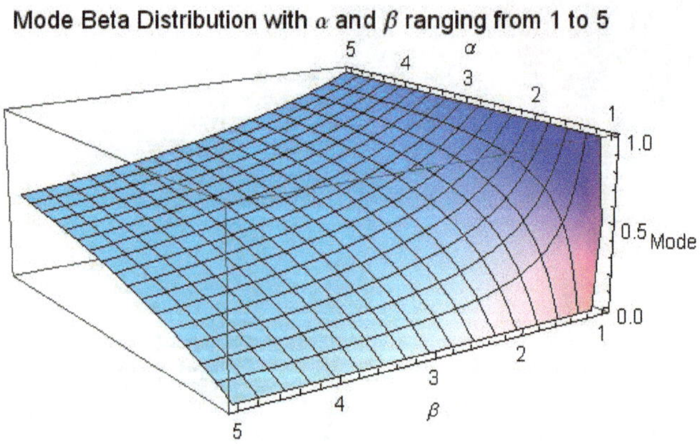

Mode for Beta distribution for 1 ≤ α ≤ 5 and 1 ≤ β ≤ 5

- Whether the ends are part of the domain of the density function
- Whether a singularity can ever be called a mode
- Whether cases with two maxima should be called bimodal

Median

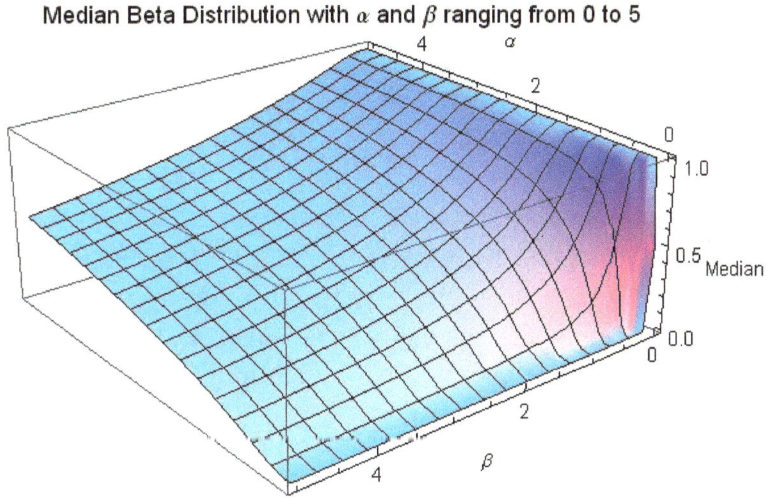

Median for Beta distribution for 0≤α≤5 and 0≤β≤5.

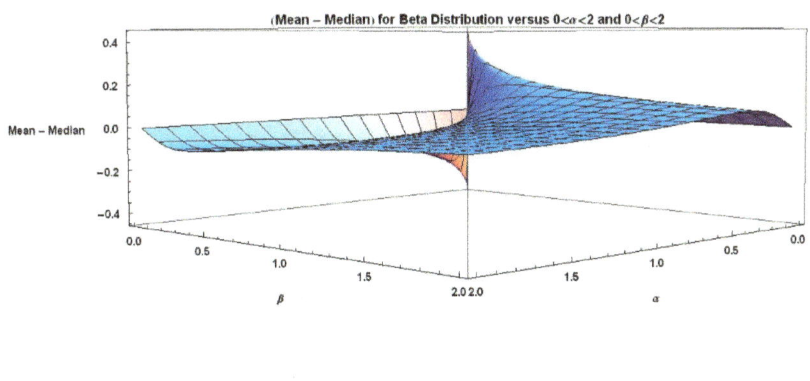

(Mean - Median) for Beta distribution versus alpha and beta from 0 to 2.

The median of the beta distribution is the unique real number $x = I_{\frac{1}{2}}^{[-1]}(\alpha, \beta)$ for which the regularized incomplete beta function $I_x(\alpha, \beta) = \frac{1}{2}$. There is no general closed-form expression for the median of the beta distribution for arbitrary values of α and β. Closed-form expressions for particular values of the parameters α and β follow:

- For symmetric cases α = β, median = 1/2.
- For α = 1 and β > 0, median = $1-2^{-\frac{1}{\beta}}$ (this case is the mirror-image of the power function [0,1] distribution)
- For α > 0 and β = 1, median = $2^{-\frac{1}{\alpha}}$ (this case is the power function [0,1] distribution)
- For α = 3 and β = 2, median = 0.6142724318676105..., the real solution to the quartic equation $1-8x^3+6x^4 = 0$, which lies in [0,1].
- For α = 2 and β = 3, median = 0.38572756813238945... = 1−median(Beta(3, 2))

The following are the limits with one parameter finite (non zero) and the other approaching these limits:

$$\lim_{\beta\to 0} \text{median} = \lim_{\alpha\to\infty} \text{median} = 1,$$

$$\lim_{\alpha\to 0} \text{median} = \lim_{\beta\to\infty} \text{median} = 0.$$

A reasonable approximation of the value of the median of the beta distribution, for both α and β greater or equal to one, is given by the formula

$$\text{median} \approx \frac{\alpha - \frac{1}{3}}{\alpha + \beta - \frac{2}{3}} \text{ for } \alpha, \beta \geq 1.$$

When α, β ≥ 1, the relative error (the absolute error divided by the median) in this approximation is less than 4% and for both α ≥ 2 and β ≥ 2 it is less than 1%. The absolute error divided by the difference between the mean and the mode is similarly small:

Mean

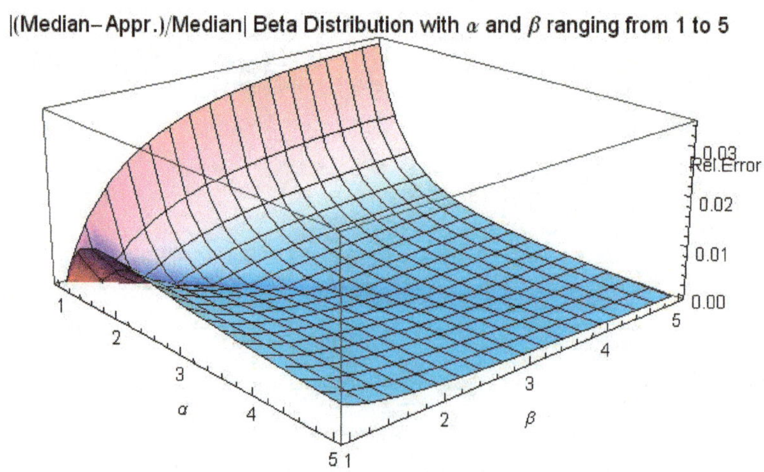

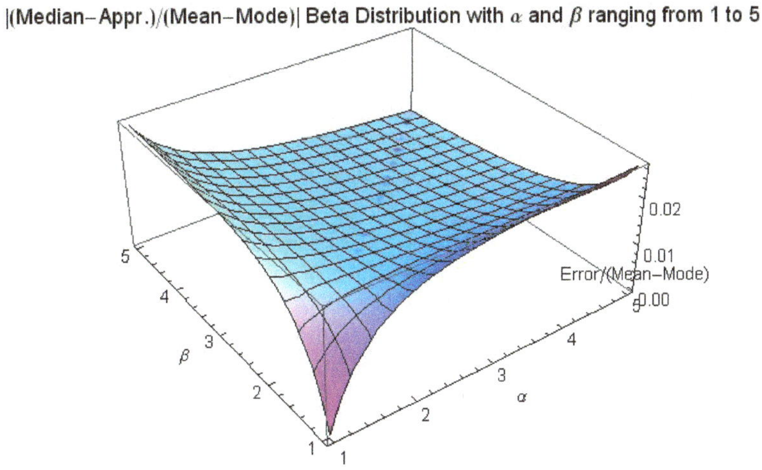

Mean for Beta distribution for $0 \leq \alpha \leq 5$ and $0 \leq \beta \leq 5$.

The expected value (mean) (μ) of a Beta distribution random variableX with two parameters α and β is a function of only the ratio β/α of these parameters:

$$\begin{aligned}\mu = \mathrm{E}[X] &= \int_0^1 x f(x;\alpha,\beta)\,dx \\ &= \int_0^1 x \frac{x^{\alpha-1}(1-x)^{\beta-1}}{\mathrm{B}(\alpha,\beta)}\,dx \\ &= \frac{\alpha}{\alpha+\beta} \\ &= \frac{1}{1+\frac{\beta}{\alpha}}\end{aligned}$$

Letting $\alpha = \beta$ in the above expression one obtains μ = 1/2, showing that for α = β the mean is at the center of the distribution: it is symmetric. Also, the following limits can be obtained from the above expression:

$$\lim_{\frac{\beta}{\alpha}\to 0} \mu = 1$$

$$\lim_{\frac{\beta}{\alpha}\to \infty} \mu = 0$$

Therefore, for $\beta/\alpha \rightarrow 0$, or for $\alpha/\beta \rightarrow \infty$, the mean is located at the right end, x = 1. For these limit ratios, the beta distribution becomes a one-point degenerate distribution with a Dirac delta function spike at the right end, x = 1, with probability 1, and zero probability everywhere else. There is 100% probability (absolute certainty) concentrated at the right end, x = 1.

Similarly, for β/α → ∞, or for α/β → 0, the mean is located at the left end, x = 0. The beta distribution becomes a 1-point Degenerate distribution with a Dirac delta function spike at the left end, x = 0, with probability 1, and zero probability everywhere else. There is 100% probability (absolute certainty) concentrated at the left end, x = 0.

Following are the limits with one parameter finite (non zero) and the other approaching these limits:

$$\lim_{\beta\to 0}\mu = \lim_{\alpha\to\infty}\mu = 1$$

$$\lim_{\alpha\to 0}\mu = \lim_{\beta\to\infty}\mu = 0$$

While for typical unimodal distributions (with centrally located modes, inflexion points at both sides of the mode, and longer tails) (with Beta(α, β) such that α, β > 2) it is known that the sample mean (as an estimate of location) is not as robust as the sample median, the opposite is the case for uniform or "U-shaped" bimodal distributions (with Beta(α, β) such that α, β ≤ 1), with the modes located at the ends of the distribution. As Mosteller and Tukey remark (p. 207) «the average of the two extreme observations uses all the sample information. This illustrates how, for short-tailed distributions, the extreme observations should get more weight.» By contrast, it follows that the median of «U-shaped» bimodal distributions with modes at the edge of the distribution (with Beta(α, β) such that α, β ≤ 1) is not robust, as the sample median drops the extreme sample observations from consideration. A practical application of this occurs for example for random walks, since the probability for the time of the last visit to the origin in a random walk is distributed as the arcsine distribution Beta(1/2, 1/2): the mean of a number of realizations of a random walk is a much more robust estimator than the median (which is an inappropriate sample measure estimate in this case).

Geometric mean

(Mean − GeometricMean) for Beta distribution versus α and β from 0 to 2, showing the asymmetry between α and β for the geometric mean

Geometric means for Beta distribution Purple = G(x), Yellow = G(1 − x), smaller values alpha and beta in front

Geometric means for Beta distribution. purple = G(x), yellow = G(1 − x), larger values α and β in front

The logarithm of the geometric mean G_X of a distribution with random variable X is the arithmetic mean of ln(X), or, equivalently, its expected value:

$$\ln G_X = \mathrm{E}[\ln X]$$

For a beta distribution, the expected value integral gives:

Therefore, the geometric mean of a beta distribution with shape parameters α and β is the exponential of the digamma functions of α and β as follows:

$$G_X = e^{\mathrm{E}[\ln X]} = e^{\psi(\alpha)-\psi(\alpha+\beta)}$$

While for a beta distribution with equal shape parameters α = β, it follows that skewness = 0 and mode = mean = median = 1/2, the geometric mean is less than 1/2: $0 < G_X < 1/2$. The reason for this is that the logarithmic transformation strongly weights the values of X close to zero, as ln(X) strongly tends towards negative infinity as X approaches zero, while ln(X) flattens towards zero as X → 1.

$$
\begin{aligned}
\mathrm{E}[\ln X] &= \int_0^1 \ln x\, f(x;\alpha,\beta)\, dx \\
&= \int_0^1 \ln x \frac{x^{\alpha-1}(1-x)^{\beta-1}}{\mathrm{B}(\alpha,\beta)}\, dx \\
&= \frac{1}{\mathrm{B}(\alpha,\beta)} \int_0^1 \frac{\partial x^{\alpha-1}(1-x)^{\beta-1}}{\partial\alpha}\, dx \\
&= \frac{1}{\mathrm{B}(\alpha,\beta)} \frac{\partial}{\partial\alpha} \int_0^1 x^{\alpha-1}(1-x)^{\beta-1}\, dx \\
&= \frac{1}{\mathrm{B}(\alpha,\beta)} \frac{\partial \mathrm{B}(\alpha,\beta)}{\partial\alpha} \\
&= \frac{\partial \ln \mathrm{B}(\alpha,\beta)}{\partial\alpha} \\
&= \frac{\partial \ln \Gamma(\alpha)}{\partial\alpha} - \frac{\partial \ln \Gamma(\alpha+\beta)}{\partial\alpha} \\
&= \psi(\alpha) - \psi(\alpha+\beta)
\end{aligned}
$$

where ψ is the digamma function.

Along a line α = β, the following limits apply:

$$\lim_{\alpha=\beta\to 0} G_X = 0$$

$$\lim_{\alpha=\beta\to\infty} G_X = \tfrac{1}{2}$$

Following are the limits with one parameter finite (non zero) and the other approaching these limits:

$$\lim_{\beta\to 0} G_X = \lim_{\alpha\to\infty} G_X = 1$$

$$\lim_{\alpha\to 0} G_X = \lim_{\beta\to\infty} G_X = 0$$

The accompanying plot shows the difference between the mean and the geometric mean for shape parameters α and β from zero to 2. Besides the fact that the difference between them approaches zero as α and β approach infinity and that the difference becomes large for values of α and β approaching zero, one can observe an evident asymmetry of the geometric mean with respect to the shape parameters α and β. The difference between the geometric mean and the mean is larger for small values of α in relation to β than when exchanging the magnitudes of β and α.

N. L.Johnson and S. Kotz suggest the logarithmic approximation to the digamma function ψ(α) ≈ ln(α-1/2) which results in the following approximation to the geometric mean:

$$G_X \approx \frac{\alpha - \frac{1}{2}}{\alpha + \beta - \frac{1}{2}} \text{ if } \alpha, \beta > 1.$$

Numerical values for the relative error in this approximation follow: [(α = β = 1): 9.39%]; [(α = β = 2): 1.29%]; [(α = 2, β = 3): 1.51%]; [(α = 3, β = 2): 0.44%]; [(α = β = 3): 0.51%]; [(α = β = 4): 0.26%]; [(α = 3, β = 4): 0.55%]; [(α = 4, β = 3): 0.24%].

Similarly, one can calculate the value of shape parameters required for the geometric mean to equal 1/2. Given the value of the parameter β, what would be the value of the other parameter, α, required for the geometric mean to equal 1/2?. The answer is that (for β > 1), the value of α required tends towards β + 1/2 as β → ∞. For example, all these couples have the same geometric mean of 1/2: [β = 1, α = 1.4427], [β = 2, α = 2.46958], [β = 3, α = 3.47943], [β = 4, α = 4.48449], [β = 5, α = 5.48756], [β = 10, α = 10.4938], [β = 100, α = 100.499].

The fundamental property of the geometric mean, which can be proven to be false for any other mean, is

$$G\left(\frac{X_i}{Y_i}\right) = \frac{G(X_i)}{G(Y_i)}$$

This makes the geometric mean the only correct mean when averaging normalized results, that is results that are presented as ratios to reference values. This is relevant because the beta distribution is a suitable model for the random behavior of percentages and it is particularly suitable to the statistical modelling of proportions. The geometric mean plays a central role in maximum likelihood estimation, see section "Parameter estimation, maximum likelihood." Actually, when performing maximum likelihood estimation, besides the geometric mean G_X based on the random variable X, also another geometric mean appears naturally: the geometric mean based on the linear transformation —(1 − X), the mirror-image of X, denoted by $G_{(1-X)}$:

$$G_{(1-X)} = e^{\mathrm{E}[\ln(1-X)]} = e^{\psi(\beta)-\psi(\alpha+\beta)}$$

Along a line α = β, the following limits apply:

$$\lim_{\alpha=\beta\to 0} G_{(1-X)} = 0$$

$$\lim_{\alpha=\beta\to\infty} G_{(1-X)} = \frac{1}{2}$$

Following are the limits with one parameter finite (non-zero) and the other approaching these limits:

$$\lim_{\beta \to 0} G_{(1-X)} = \lim_{\alpha \to \infty} G_{(1-X)} = 0$$

$$\lim_{\alpha \to 0} G_{(1-X)} = \lim_{\beta \to \infty} G_{(1-X)} = 1$$

It has the following approximate value:

$$G_{(1-X)} \approx \frac{\beta - \frac{1}{2}}{\alpha + \beta - \frac{1}{2}} \text{ if } \alpha, \beta > 1.$$

Although both G_X and $G_{(1-X)}$ are asymmetric, in the case that both shape parameters are equal $\alpha = \beta$, the geometric means are equal: $G_X = G_{(1-X)}$. This equality follows from the following symmetry displayed between both geometric means:

$$G_X(\mathrm{B}(\alpha, \beta)) = G_{(1-X)}(\mathrm{B}(\beta, \alpha)).$$

Harmonic Mean

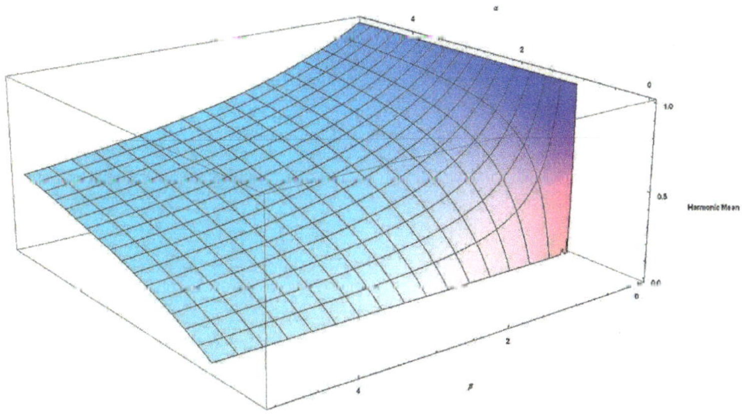

Harmonic mean for Beta distribution for 0<α<5 and 0<β<5.

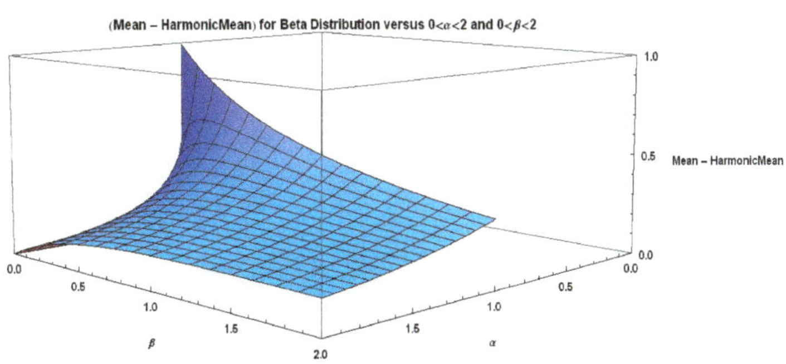

(Mean - HarmonicMean) for Beta distribution versus alpha and beta from 0 to 2.

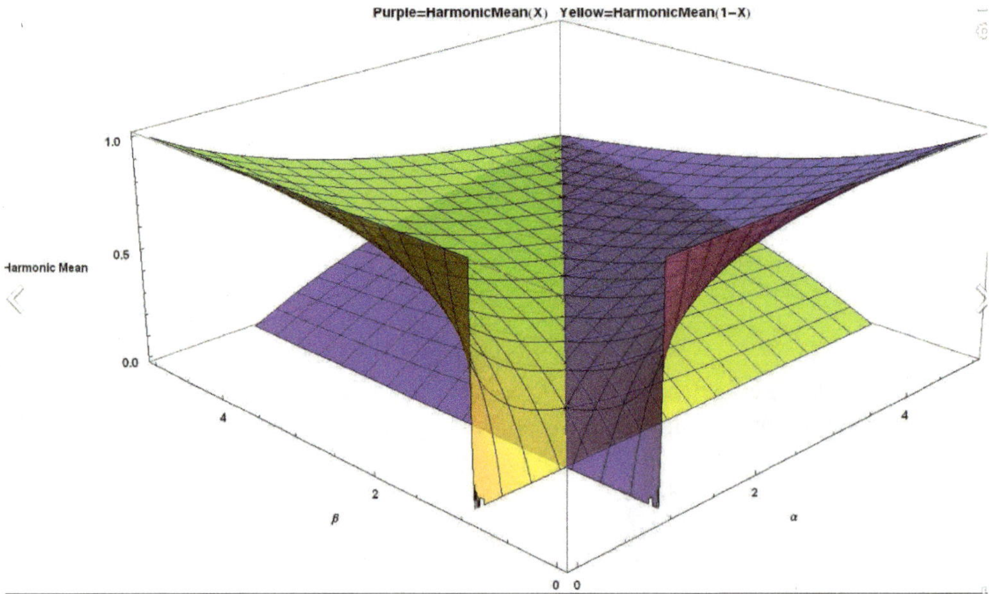

Harmonic Means for Beta distribution Purple=H(X), Yellow=H(1-X), smaller values alpha and beta in front.

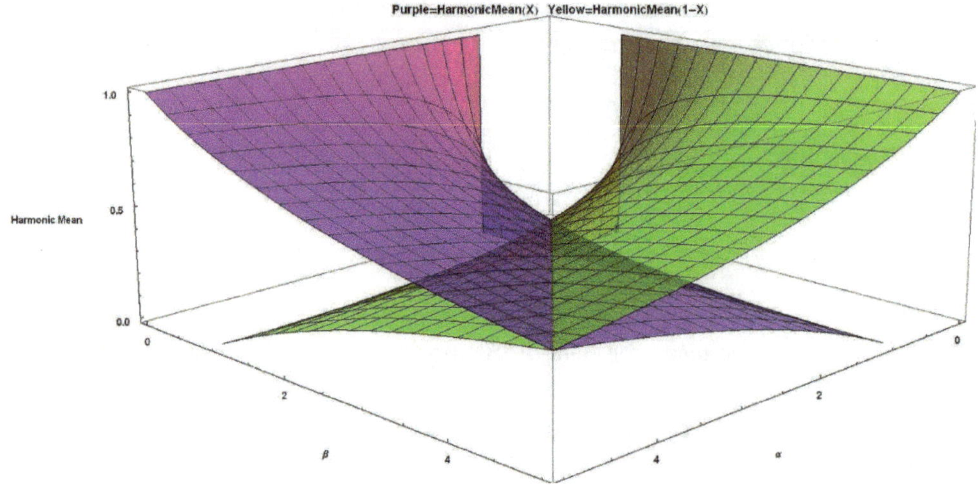

Harmonic Means for Beta distribution Purple=H(X), Yellow=H(1-X), larger values alpha and beta in front

The inverse of the harmonic mean (H_X) of a distribution with random variable X is the arithmetic mean of 1/X, or, equivalently, its expected value. Therefore, the harmonic mean (H_X) of a beta distribution with shape parameters α and β is:

$$
\begin{aligned}
H_X &= \frac{1}{\mathrm{E}\left[\frac{1}{X}\right]} \\
&= \frac{1}{\int_0^1 \frac{f(x;\alpha,\beta)}{x}\, dx} \\
&= \frac{1}{\int_0^1 \frac{x^{\alpha-1}(1-x)^{\beta-1}}{x\mathrm{B}(\alpha,\beta)}\, dx} \\
&= \frac{\alpha-1}{\alpha+\beta-1} \text{ if } \alpha > 1 \text{ and } \beta > 0
\end{aligned}
$$

The harmonic mean (H_X) of a Beta distribution with $\alpha < 1$ is undefined, because its defining expression is not bounded in [0, 1] for shape parameter α less than unity.

Letting $\alpha = \beta$ in the above expression one obtains

$$G_X = e^{\mathrm{E}[\ln X]} = e^{\psi(\alpha)-\psi(\alpha+\beta)}$$

showing that for $\alpha = \beta$ the harmonic mean ranges from 0, for $\alpha = \beta = 1$, to 1/2, for $\alpha = \beta \to \infty$.

Following are the limits with one parameter finite (non zero) and the other approaching these limits:

$$\lim_{\alpha=\beta\to 0} G_X = 0$$

$$\lim_{\alpha=\beta\to\infty} G_X = \tfrac{1}{2}$$

The harmonic mean plays a role in maximum likelihood estimation for the four parameter case, in addition to the geometric mean. Actually, when performing maximum likelihood estimation for the four parameter case, besides the harmonic mean H_X based on the random variable X, also another harmonic mean appears naturally: the harmonic mean based on the linear transformation (1–X), the mirror-image of X, denoted by $H_{(1-X)}$:

$$G_{(1-X)} = e^{\mathrm{E}[\ln(1-X)]} = e^{\psi(\beta)-\psi(\alpha+\beta)}$$

Along a line $\alpha = \beta$, the following limits apply:

$$\lim_{\alpha=\beta\to 0} G_{(1-X)} = 0$$

$$\lim_{\alpha=\beta\to\infty} G_{(1-X)} = \frac{1}{2}$$

Following are the limits with one parameter finite (non-zero) and the other approaching these limits:

$$\lim_{\beta\to 0} G_{(1-X)} = \lim_{\alpha\to\infty} G_{(1-X)} = 0$$

$$\lim_{\alpha\to 0} G_{(1-X)} = \lim_{\beta\to\infty} G_{(1-X)} = 1$$

It has the following approximate value:

$$G_{(1-X)} \approx \frac{\beta - \frac{1}{2}}{\alpha+\beta-\frac{1}{2}} \text{ if } \alpha, \beta > 1.$$

Although both H_X and $H_{(1-X)}$ are asymmetric, in the case that both shape parameters are equal α = β, the harmonic means are equal: $H_X = H_{(1-X)}$. This equality follows from the following symmetry displayed between both harmonic means:

$$G_X(\mathrm{B}(\alpha,\beta)) = G_{(1-X)}(\mathrm{B}(\beta,\alpha)).$$

Measures of Statistical Dispersion

Variance

The variance (the second moment centered on the mean) of a Beta distribution random variable X with parameters α and β is:

$$\mathrm{var}(X) = \mathrm{E}[(X-\mu)^2] = \frac{\alpha\beta}{(\alpha+\beta)^2(\alpha+\beta+1)}$$

Letting α = β in the above expression one obtains

$$\mathrm{var}(X) = \frac{1}{4(2\beta+1)},$$

showing that for α = β the variance decreases monotonically as Template:Nowwrap increases. Setting α = β = 0 in this expression, one finds the maximum variance var(X) = 1/4 which only occurs approaching the limit, at α = β = 0.

The beta distribution may also be parametrized in terms of its mean μ (0 < μ < 1) and sample size ν = α + β (ν > 0) (see section below titled «Mean and sample size»):

$$\alpha = \mu\nu, \text{ where } \nu = (\alpha+\beta) > 0$$
$$\beta = (1-\mu)\nu, \text{ where } \nu = (\alpha+\beta) > 0.$$

Using this parametrization, one can express the variance in terms of the mean μ and the sample size ν as follows:

$$\mathrm{var}(X) = \frac{\mu(1-\mu)}{1+\nu}$$

Since {{{1}}}, it must follow that var(X) < μ(1 − μ)

For a symmetric distribution, the mean is at the middle of the distribution, {{{1}}}, and therefore:

$$\mathrm{var}(X) = \frac{1}{4(1+\nu)} \text{ if } \mu = \tfrac{1}{2}$$

Also, the following limits (with only the noted variable approaching the limit) can be obtained from the above expressions:

$$\lim_{\beta\to 0} \text{var}(X) = \lim_{\alpha\to 0} \text{var}(X) = \lim_{\beta\to\infty} \text{var}(X) = \lim_{\alpha\to\infty} \text{var}(X) = \lim_{\nu\to\infty} \text{var}(X) = \lim_{\mu\to 0} \text{var}(X) = \lim_{\mu\to 1} \text{var}(X) = 0$$

$$\lim_{\nu\to 0} \text{var}(X) = \mu(1-\mu)$$

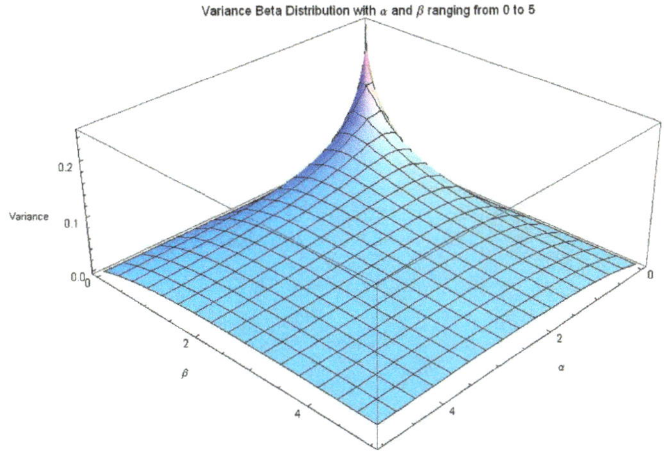

Geometric Variance and Covariance

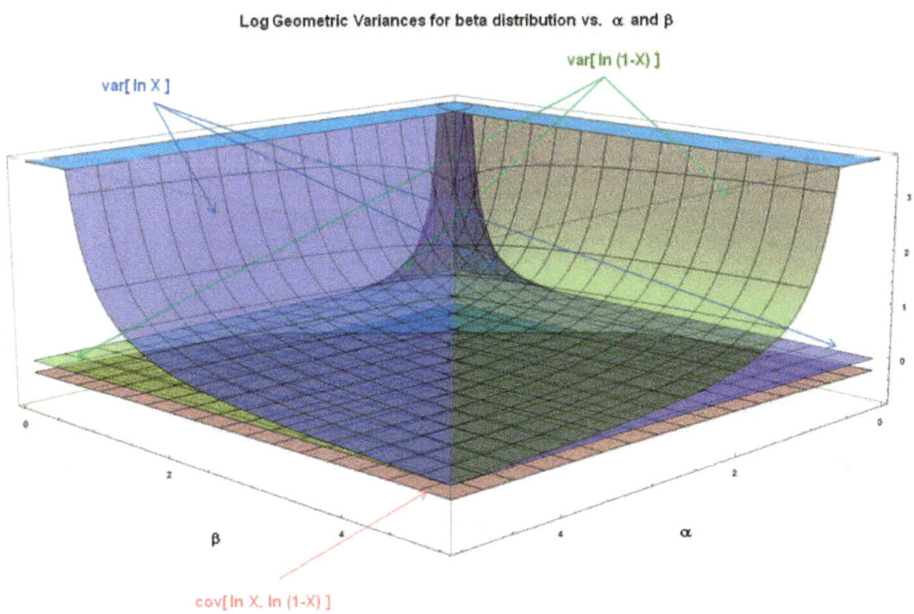

log geometric variances vs. α and β

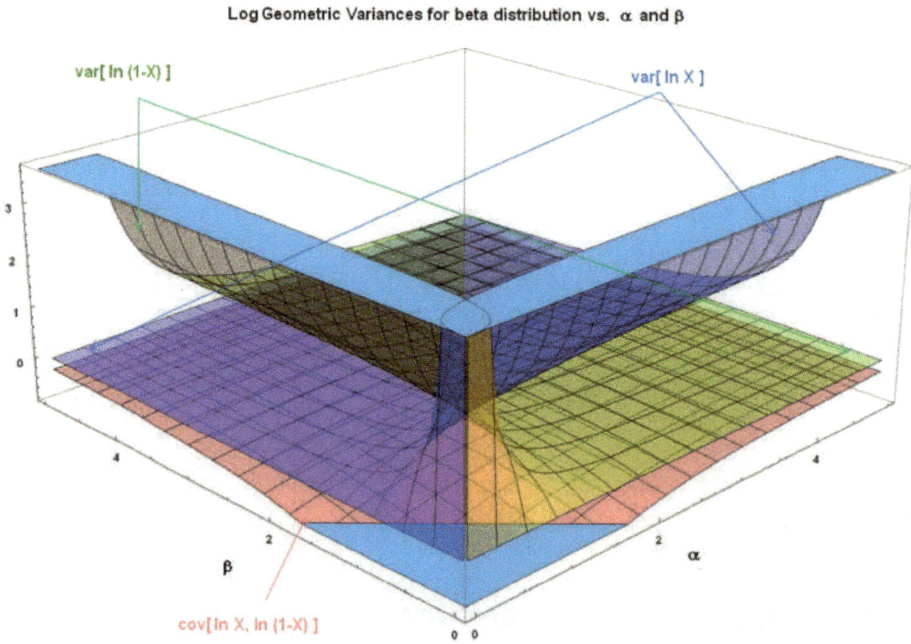

log geometric variances vs. α and β

The logarithm of the geometric variance, $\ln(\mathrm{var}_{GX})$, of a distribution with random variable X is the second moment of the logarithm of X centered on the geometric mean of X, $(\ln(G_X)$:

$$\begin{aligned}\ln \mathrm{var}_{GX} &= \mathrm{E}\left[(\ln X - \ln G_X)^2\right] \\ &= \mathrm{E}[(\ln X - \mathrm{E}[\ln X])^2] \\ &= \mathrm{E}\left[(\ln X)^2\right] - (\mathrm{E}[\ln X])^2 \\ &= \mathrm{var}[\ln X]\end{aligned}$$

and therefore, the geometric variance is:

$$\mathrm{var}_{GX} = e^{\mathrm{var}[\ln X]}$$

In the Fisher information matrix, and the curvature of the log likelihood function, the logarithm of the geometric variance of the reflected variable (1-X) and the logarithm of the geometric covariance between X and (1-X) appear:

$$\begin{aligned}\ln \mathrm{var}_{G(1-X)} &= \mathrm{E}[(\ln(1-X) - \ln G_{(1-X)})^2] \\ &= \mathrm{E}[(\ln(1-X) - \mathrm{E}[\ln(1-X)])^2] \\ &= \mathrm{E}[(\ln(1-X))^2] - (\mathrm{E}[\ln(1-X)])^2 \\ &= \mathrm{var}[\ln(1-X)]\end{aligned}$$

$$\mathrm{var}_{G(1-X)} = e^{\mathrm{var}[\ln(1-X)]}$$

$$\begin{aligned}\ln \mathrm{cov}_{GX,(1-X)} &= \mathrm{E}[(\ln X - \ln G_X)(\ln(1-X) - \ln G_{(1-X)})] \\ &= \mathrm{E}[(\ln X - \mathrm{E}[\ln X])(\ln(1-X) - \mathrm{E}[\ln(1-X)])] \\ &= \mathrm{E}[\ln X \ln(1-X)] - \mathrm{E}[\ln X]\,\mathrm{E}[\ln(1-X)] \\ &= \mathrm{cov}[\ln X, \ln(1-X)]\end{aligned}$$

$$\mathrm{cov}_{GX,(1-X)} = e^{\mathrm{cov}[\ln X, \ln(1-X)]}$$

For a beta distribution, higher order logarithmic moments can be derived by using the representation of a beta distribution as a proportion of two Gamma distributions and differentiating through the integral. They can be expressed in terms of higher order polygamma functions. See the section titled "Other moments, Moments of transformed random variables, Moments of logarithmically transformed random variables". The variance of the logarithmic variables and covariance of lnX and ln(1−X) are:

$$\begin{aligned}\mathrm{var}[\ln X] &= \psi_1(\alpha) - \psi_1(\alpha+\beta) \\ \mathrm{var}[\ln(1-X)] &= \psi_1(\beta) - \psi_1(\alpha+\beta) \\ \mathrm{cov}[\ln X, \ln(1-X)] &= -\psi_1(\alpha+\beta)\end{aligned}$$

where the **trigamma function**, denoted $\psi_1(\alpha)$, is the second of the polygamma functions, and is defined as the derivative of the digamma function:

$$\psi_1(\alpha) = \frac{d^2 \ln\Gamma(\alpha)}{d\alpha^2} = \frac{d\,\psi(\alpha)}{d\alpha}.$$

Therefore,

$$\begin{aligned}\ln \mathrm{var}_{GX} &= \mathrm{var}[\ln X] = \psi_1(\alpha) - \psi_1(\alpha+\beta) \\ \ln \mathrm{var}_{G(1-X)} &= \mathrm{var}[\ln(1-X)] = \psi_1(\beta) - \psi_1(\alpha+\beta) \\ \ln \mathrm{cov}_{GX,(1-X)} &= \mathrm{cov}[\ln X, \ln(1-X)] = -\psi_1(\alpha+\beta)\end{aligned}$$

The accompanying plots show the log geometric variances and log geometric covariance versus the shape parameters α and β. The plots show that the log geometric variances and log geometric covariance are close to zero for shape parameters α and β greater than 2, and that the log geometric variances rapidly rise in value for shape parameter values α and β less than unity. The log geometric variances are positive for all values of the shape parameters. The log geometric covariance is negative for all values of the shape parameters, and it reaches large negative values for α and β less than unity.

Following are the limits with one parameter finite (non zero) and the other approaching these limits:

$$\lim_{\alpha\to 0} \ln \text{var}_{GX} = \lim_{\beta\to 0} \ln \text{var}_{G(1-X)} = \infty$$

$$\lim_{\beta\to 0} \ln \text{var}_{GX} = \lim_{\alpha\to\infty} \ln \text{var}_{GX} = \lim_{\alpha\to 0} \ln \text{var}_{G(1-X)} = \lim_{\beta\to\infty} \ln \text{var}_{G(1-X)} = \lim_{\alpha\to\infty} \ln \text{cov}_{GX,(1-X)} = \lim_{\beta\to\infty} \ln \text{cov}_{GX,(1-X)} = 0$$

$$\lim_{\beta\to\infty} \ln \text{var}_{GX} = \psi_1(\alpha)$$

$$\lim_{\alpha\to\infty} \ln \text{var}_{G(1-X)} = \psi_1(\beta)$$

$$\lim_{\alpha\to 0} \ln \text{cov}_{GX,(1-X)} = -\psi_1(\beta)$$

$$\lim_{\beta\to 0} \ln \text{cov}_{GX,(1-X)} = -\psi_1(\alpha)$$

Limits with two parameters varying:

$$\lim_{\alpha\to\infty}(\lim_{\beta\to\infty} \ln \text{var}_{GX}) = \lim_{\beta\to\infty}(\lim_{\alpha\to\infty} \ln \text{var}_{G(1-X)}) = \lim_{\alpha\to\infty}(\lim_{\beta\to 0} \ln \text{cov}_{GX,(1-X)}) = \lim_{\beta\to\infty}(\lim_{\alpha\to 0} \ln \text{cov}_{GX,(1-X)}) = 0$$

$$\lim_{\alpha\to\infty}(\lim_{\beta\to 0} \ln \text{var}_{GX}) = \lim_{\beta\to\infty}(\lim_{\alpha\to 0} \ln \text{var}_{G(1-X)}) = \infty$$

$$\lim_{\alpha\to 0}(\lim_{\beta\to 0} \ln \text{cov}_{GX,(1-X)}) = \lim_{\beta\to 0}(\lim_{\alpha\to 0} \ln \text{cov}_{GX,(1-X)}) = -\infty$$

Although both $\ln(\text{var}_{GX})$ and $\ln(\text{var}_{G(1-X)})$ are asymmetric, when the shape parameters are equal, $\alpha = \beta$, one has: $\ln(\text{var}_{GX}) = \ln(\text{var}_{G(1-X)})$. This equality follows from the following symmetry displayed between both log geometric variances:

$$\ln \text{var}_{GX}(\text{B}(\alpha,\beta)) = \ln \text{var}_{G(1-X)}(\text{B}(\beta,\alpha)).$$

The log geometric covariance is symmetric:

$$\ln \text{cov}_{GX,(1-X)}(\text{B}(\alpha,\beta)) = \ln \text{cov}_{GX,(1-X)}(\text{B}(\beta,\alpha))$$

Mean Absolute Deviation Around the Mean

Ratio of Mean Abs.Dev. to Std.Dev. for Beta distribution with α and β ranging from 0 to 5

Ratio of Mean Abs.Dev. to Std.Dev. for Beta distribution with mean $0 \le \mu \le 1$ and sample size $0 < \nu \le 10$

The mean absolute deviation around the mean for the beta distribution with shape parameters α and β is:

$$\text{E}[|X - E[X]|] = \frac{2\alpha^\alpha \beta^\beta}{\text{B}(\alpha,\beta)(\alpha+\beta)^{\alpha+\beta+1}}$$

The mean absolute deviation around the mean is a more robust estimator of statistical dispersion than the standard deviation for beta distributions with tails and inflection points at each side of the mode, Beta(α, β) distributions with $\alpha,\beta > 2$, as it depends on the linear (absolute) deviations rather than the square deviations from the mean. Therefore, the effect of very large deviations from the mean are not as overly weighted.

Using Stirling›s approximation to the Gamma function, N.L.Johnson and S.Kotzderived the following approximation for values of the shape parameters greater than unity (the relative error for this approximation is only −3.5% for $\alpha = \beta = 1$, and it decreases to zero as $\alpha \to \infty$, $\beta \to \infty$):

$$\frac{\text{mean abs. dev. from mean}}{\text{standard deviation}} = \frac{\mathrm{E}[|X - E[X]|]}{\sqrt{\mathrm{var}(X)}}$$
$$\approx \sqrt{\frac{2}{\pi}}\left(1 + \frac{7}{12(\alpha+\beta)} - \frac{1}{12\alpha} - \frac{1}{12\beta}\right), \text{ if } \alpha, \beta > 1.$$

At the limit $\alpha \to \infty$, $\beta \to \infty$, the ratio of the mean absolute deviation to the standard deviation (for the beta distribution) becomes equal to the ratio of the same measures for the normal distribution: $\sqrt{\frac{2}{\pi}}$. For $\alpha = \beta = 1$ this ratio equals $\frac{\sqrt{3}}{2}$, so that from $\alpha = \beta = 1$ to α, $\beta \to \infty$ the ratio decreases by 8.5%. For $\alpha = \beta = 0$ the standard deviation is exactly equal to the mean absolute deviation around the mean. Therefore, this ratio decreases by 15% from $\alpha = \beta = 0$ to $\alpha = \beta = 1$, and by 25% from $\alpha = \beta = 0$ to α, $\beta \to \infty$. However, for skewed beta distributions such that $\alpha \to 0$ or $\beta \to 0$, the ratio of the standard deviation to the mean absolute deviation approaches infinity (although each of them, individually, approaches zero) because the mean absolute deviation approaches zero faster than the standard deviation.

Using the parametrization in terms of mean μ and sample size $\nu = \alpha + \beta > 0$:

$\alpha = \mu\nu$, $\beta = (1-\mu)\nu$

one can express the mean absolute deviation around the mean in terms of the mean μ and the sample size ν as follows:

$$\mathrm{E}[|X - E[X]|] = \frac{2\mu^{\mu\nu}(1-\mu)^{(1-\mu)\nu}}{\nu \mathrm{B}(\mu\nu, (1-\mu)\nu)}$$

For a symmetric distribution, the mean is at the middle of the distribution, $\mu = 1/2$, and therefore:

$$\mathrm{E}[|X - E[X]|] = \frac{2^{1-\nu}}{\nu \mathrm{B}(\frac{\nu}{2}, \frac{\nu}{2})} = \frac{2^{1-\nu}\Gamma(\nu)}{\nu(\Gamma(\frac{\nu}{2}))^2}$$

$$\lim_{\nu \to 0}\left(\lim_{\mu \to \frac{1}{2}} \mathrm{E}[|X - E[X]|]\right) = \frac{1}{2}$$

$$\lim_{\nu \to \infty}\left(\lim_{\mu \to \frac{1}{2}} \mathrm{E}[|X - E[X]|]\right) = 0$$

Also, the following limits (with only the noted variable approaching the limit) can be obtained from the above expressions:

$$\lim_{\beta \to 0} \mathrm{E}[|X - E[X]|] = \lim_{\alpha \to 0} \mathrm{E}[|X - E[X]|] = 0$$
$$\lim_{\beta \to \infty} \mathrm{E}[|X - E[X]|] = \lim_{\alpha \to \infty} \mathrm{E}[|X - E[X]|] = 0$$
$$\lim_{\mu \to 0} \mathrm{E}[|X - E[X]|] = \lim_{\mu \to 1} \mathrm{E}[|X - E[X]|] = 0$$
$$\lim_{\nu \to 0} \mathrm{E}[|X - E[X]|] = \sqrt{\mu(1-\mu)}$$
$$\lim_{\nu \to \infty} \mathrm{E}[|X - E[X]|] = 0$$

Mean absolute difference

The mean absolute difference for the Beta distribution is:

$$\mathrm{MD} = \int_0^1 \int_0^1 f(x;\alpha,\beta)\, f(y;\alpha,\beta)\, |x-y|\, dx\, dy = \left(\frac{4}{\alpha+\beta}\right) \frac{B(\alpha+\beta,\alpha+\beta)}{B(\alpha,\alpha)B(\beta,\beta)}$$

The Gini coefficient for the Beta distribution is half of the relative mean absolute difference:

$$\mathrm{G} = \left(\frac{2}{\alpha}\right) \frac{B(\alpha+\beta,\alpha+\beta)}{B(\alpha,\alpha)B(\beta,\beta)}$$

Skewness

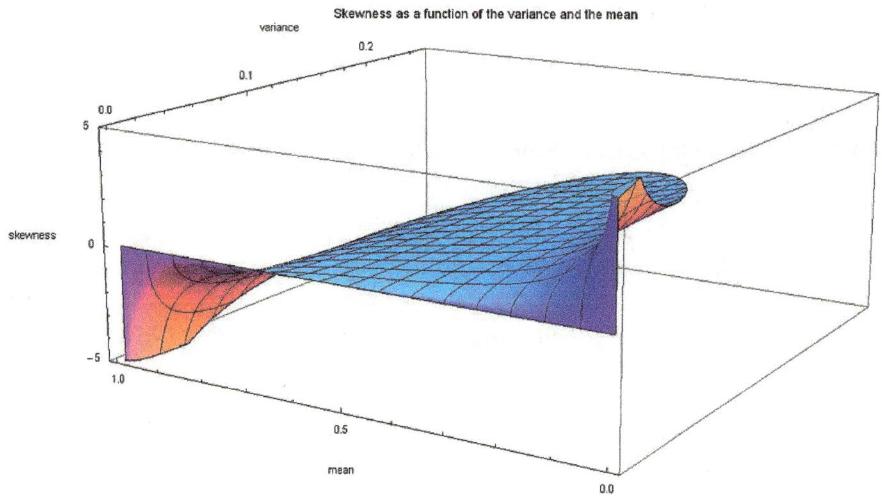

Skewness for Beta Distribution as a function of variance and mean.

The skewness (the third moment centered on the mean, normalized by the 3/2 power of the variance) of the beta distribution is

$$\gamma_1 = \frac{\mathrm{E}[(X-\mu)^3]}{(\mathrm{var}(X))^{3/2}} = \frac{2(\beta-\alpha)\sqrt{\alpha+\beta+1}}{(\alpha+\beta+2)\sqrt{\alpha\beta}}.$$

Letting $\alpha = \beta$ in the above expression one obtains $\gamma_1 = 0$, showing once again that for $\alpha = \beta$ the distribution is symmetric and hence the skewness is zero. Positive skew (right-tailed) for $\alpha < \beta$, negative skew (left-tailed) for $\alpha > \beta$.

Using the parametrization in terms of mean μ and sample size $\nu = \alpha + \beta$:

$\alpha = \mu\nu$, where $\nu = (\alpha + \beta) > 0$
$\beta = (1 - \mu)\nu$, where $\nu = (\alpha + \beta) > 0.$

one can express the skewness in terms of the mean μ and the sample size ν as follows:

$$\gamma_1 = \frac{\mathrm{E}[(X-\mu)^3]}{(\mathrm{var}(X))^{3/2}} = \frac{2(1-2\mu)\sqrt{1+\nu}}{(2+\nu)\sqrt{\mu(1-\mu)}}.$$

The skewness can also be expressed just in terms of the variance var and the mean μ as follows:

$$\gamma_1 = \frac{\mathrm{E}[(X-\mu)^3]}{(\mathrm{var}(X))^{3/2}} = \frac{2(1-2\mu)\sqrt{\mathrm{var}}}{\mu(1-\mu)+\mathrm{var}} \text{ if var} < \mu(1-\mu)$$

The accompanying plot of skewness as a function of variance and mean shows that maximum variance (1/4) is coupled with zero skewness and the symmetry condition (μ = 1/2), and that maximum skewness (positive or negative infinity) occurs when the mean is located at one end or the other, so that the "mass" of the probability distribution is concentrated at the ends (minimum variance).

The following expression for the square of the skewness, in terms of the sample size ν = α + β and the variance var, is useful for the method of moments estimation of four parameters:

$$(\gamma_1)^2 = \frac{(\mathrm{E}[(X-\mu)^3])^2}{(\mathrm{var}(X))^3} = \frac{4}{(2+\nu)^2}\left(\frac{1}{\mathrm{var}} - 4(1+\nu)\right)$$

This expression correctly gives a skewness of zero for α = β, since in that case (see section titled "Variance"):

$$\mathrm{var} = \frac{1}{4(1+\nu)}.$$

For the symmetric case (α = β), skewness = 0 over the whole range, and the following limits apply:

$$\lim_{\alpha=\beta\to 0}\gamma_1 = \lim_{\alpha=\beta\to\infty}\gamma_1 = \lim_{\nu\to 0}\gamma_1 = \lim_{\nu\to\infty}\gamma_1 = \lim_{\mu\to\frac{1}{2}}\gamma_1 = 0$$

For the asymmetric cases (α ≠ β) the following limits (with only the noted variable approaching the limit) can be obtained from the above expressions:

$$\lim_{\alpha\to 0}\gamma_1 = \lim_{\mu\to 0}\gamma_1 = \infty$$

$$\lim_{\beta\to 0}\gamma_1 = \lim_{\mu\to 1}\gamma_1 = -\infty$$

$$\lim_{\alpha\to\infty}\gamma_1 = -\frac{2}{\beta}, \quad \lim_{\beta\to 0}(\lim_{\alpha\to\infty}\gamma_1) = -\infty, \quad \lim_{\beta\to\infty}(\lim_{\alpha\to\infty}\gamma_1) = 0$$

$$\lim_{\beta\to\infty}\gamma_1 = \frac{2}{\alpha}, \quad \lim_{\alpha\to 0}(\lim_{\beta\to\infty}\gamma_1) = \infty, \quad \lim_{\alpha\to\infty}(\lim_{\beta\to\infty}\gamma_1) = 0$$

$$\lim_{\nu\to 0}\gamma_1 = \frac{1-2\mu}{\sqrt{\mu(1-\mu)}}, \quad \lim_{\mu\to 0}(\lim_{\nu\to 0}\gamma_1) = \infty, \quad \lim_{\mu\to 1}(\lim_{\nu\to 0}\gamma_1) = -\infty$$

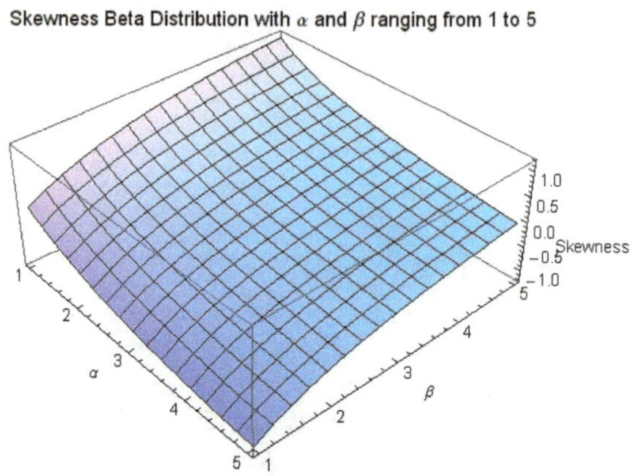

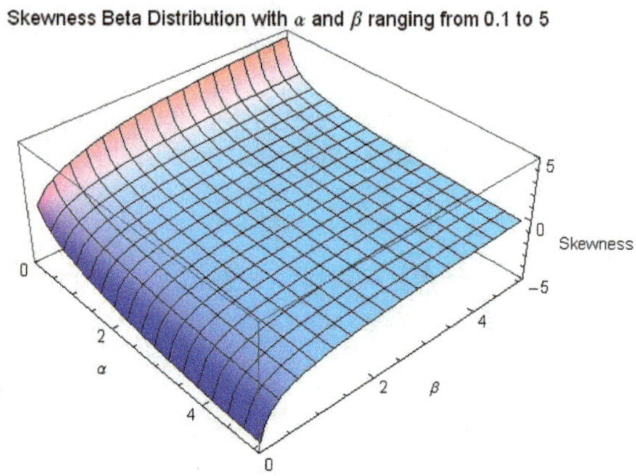

Kurtosis

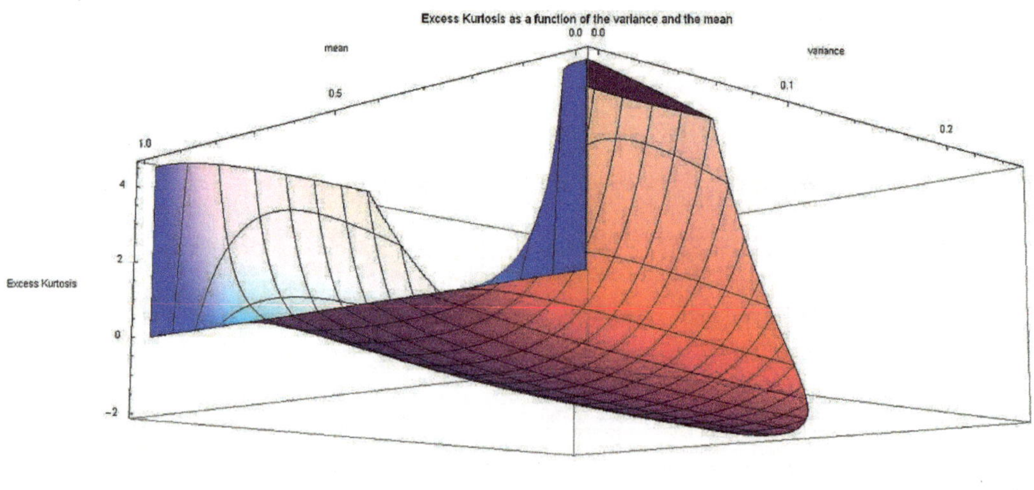

Excess Kurtosis for Beta Distribution as a function of variance and mean.

The beta distribution has been applied in acoustic analysis to assess damage to gears, as the kurtosis of the beta distribution has been reported to be a good indicator of the condition of a gear. Kurtosis has also been used to distinguish the seismic signal generated by a person's footsteps from other signals. As persons or other targets moving on the ground generate continuous signals in the form of seismic waves, one can separate different targets based on the seismic waves they generate. Kurtosis is sensitive to impulsive signals, so it's much more sensitive to the signal generated by human footsteps than other signals generated by vehicles, winds, noise, etc. Unfortunately, the notation for kurtosis has not been standardized. Kenney and Keeping use the symbol γ_2 for the excess kurtosis, but Abramowitz and Stegun use different terminology. To prevent confusion between kurtosis (the fourth moment centered on the mean, normalized by the square of the variance) and excess kurtosis, when using symbols, they will be spelled out as follows:

$$\begin{aligned}\text{excess kurtosis} &= \text{kurtosis} - 3\\ &= \frac{\mathrm{E}[(X-\mu)^4]}{(\mathrm{var}(X))^2} - 3\\ &= \frac{6[\alpha^3 - \alpha^2(2\beta - 1) + \beta^2(\beta+1) - 2\alpha\beta(\beta+2)]}{\alpha\beta(\alpha+\beta+2)(\alpha+\beta+3)}\\ &= \frac{6[(\alpha-\beta)^2(\alpha+\beta+1) - \alpha\beta(\alpha+\beta+2)]}{\alpha\beta(\alpha+\beta+2)(\alpha+\beta+3)}.\end{aligned}$$

Letting $\alpha = \beta$ in the above expression one obtains

$$\text{excess kurtosis} = -\frac{6}{3+2\alpha} \text{ if } \alpha = \beta.$$

Therefore, for symmetric beta distributions, the excess kurtosis is negative, increasing from a minimum value of −2 at the limit as $\{\alpha = \beta\} \to 0$, and approaching a maximum value of zero as $\{\alpha = \beta\} \to \infty$. The value of −2 is the minimum value of excess kurtosis that any distribution (not just beta distributions, but any distribution of any possible kind) can ever achieve. This minimum value is reached when all the probability density is entirely concentrated at each end $x = 0$ and $x = 1$, with nothing in between: a 2-point Bernoulli distribution with equal probability 1/2 at each end (a coin toss: see section below "Kurtosis bounded by the square of the skewness" for further discussion). The description of kurtosis as a measure of the "peakedness" (or "heavy tails") of the probability distribution, is strictly applicable to unimodal distributions (for example the normal distribution). However, for more general distributions, like the beta distribution, a more general description of kurtosis is that it is a measure of the proportion of the mass density near the mean. The higher the proportion of mass density near the mean, the higher the kurtosis, while the higher the mass density away from the mean, the lower the kurtosis. For $\alpha \neq \beta$, skewed beta distributions, the excess kurtosis can reach unlimited positive values (particularly for $\alpha \to 0$ for finite β, or for $\beta \to 0$ for finite α) because all the mass density is concentrated at the mean when the mean coincides with one of the ends. Minimum kurtosis takes place when the mass density is concentrated equally at each end (and therefore the mean is at the center), and there is no probability mass density in between the ends.

Using the parametrization in terms of mean μ and sample size ν = α + β:

$\alpha = \mu\nu$, where $\nu = (\alpha + \beta) > 0$
$\beta = (1 - \mu)\nu$, where $\nu = (\alpha + \beta) > 0.$

one can express the excess kurtosis in terms of the mean μ and the sample size ν as follows:

$$\text{excess kurtosis} = \frac{6}{3+\nu}\left(\frac{(1-2\mu)^2(1+\nu)}{\mu(1-\mu)(2+\nu)} - 1\right)$$

The excess kurtosis can also be expressed in terms of just the following two parameters: the variance var, and the sample size ν as follows:

$$\text{excess kurtosis} = \frac{6}{(3+\nu)(2+\nu)}\left(\frac{1}{\text{var}} - 6 - 5\nu\right) \text{ if var } < \mu(1-\mu)$$

and, in terms of the variance *var* and the mean μ as follows:

$$\text{excess kurtosis} = \frac{6 \text{ var } (1 - \text{ var } - 5\mu(1-\mu))}{(\text{var } + \mu(1-\mu))(2 \text{ var } + \mu(1-\mu))} \text{ if var } < \mu(1-\mu)$$

The plot of excess kurtosis as a function of the variance and the mean shows that the minimum value of the excess kurtosis (−2, which is the minimum possible value for excess kurtosis for any distribution) is intimately coupled with the maximum value of variance (1/4) and the symmetry condition: the mean occurring at the midpoint (μ = 1/2). This occurs for the symmetric case of α = β = 0, with zero skewness. At the limit, this is the 2 point Bernoulli distribution with equal probability 1/2 at each Dirac delta function end x = 0 and x = 1 and zero probability everywhere else. (A coin toss: one face of the coin being x = 0 and the other face being x = 1.) Variance is maximum because the distribution is bimodal with nothing in between the two modes (spikes) at each end. Excess kurtosis is minimum: the probability density "mass" is zero at the mean and it is concentrated at the two peaks at each end. Excess kurtosis reaches the minimum possible value (for any distribution) when the probability density function has two spikes at each end: it is bi-"peaky" with nothing in between them.

On the other hand, the plot shows that for extreme skewed cases, where the mean is located near one or the other end (μ = 0 or μ = 1), the variance is close to zero, and the excess kurtosis rapidly approaches infinity when the mean of the distribution approaches either end.

Alternatively, the excess kurtosis can also be expressed in terms of just the following two parameters: the square of the skewness, and the sample size ν as follows:

$$\text{excess kurtosis} = \frac{6}{3+\nu}\left(\frac{(2+\nu)}{4}(\text{skewness})^2 - 1\right) \text{ if } (\text{skewness})^2 - 2 < \text{excess kurtosis} < \frac{3}{2}(\text{skewness})^2$$

From this last expression, one can obtain the same limits published practically a century ago by Karl Pearson in his paper,for the beta distribution (see section below titled "Kurtosis bounded by the square of the skewness"). Setting α + β= ν = 0 in the above expression, one obtains Pearson's lower boundary (values for the skewness and excess kurtosis below the boundary (excess kurtosis + 2 − skewness^2 = 0) cannot occur for any distribution, and hence Karl Pearson appropriately called the region below this boundary the "impossible region"). The limit of α + β = ν → ∞ determines Pearson's upper boundary.

$$\lim_{\nu \to 0} \text{excess kurtosis} = (\text{skewness})^2 - 2$$

$$\lim_{\nu \to \infty} \text{excess kurtosis} = \tfrac{3}{2}(\text{skewness})^2$$

therefore:

$$(\text{skewness})^2 - 2 < \text{excess kurtosis} < \tfrac{3}{2}(\text{skewness})^2$$

Values of $\nu = \alpha + \beta$ such that ν ranges from zero to infinity, $0 < \nu < \infty$, span the whole region of the beta distribution in the plane of excess kurtosis versus squared skewness.

For the symmetric case ($\alpha = \beta$), the following limits apply:

$$\lim_{\alpha=\beta \to 0} \text{excess kurtosis} = -2$$

$$\lim_{\alpha=\beta \to \infty} \text{excess kurtosis} = 0$$

$$\lim_{\mu \to \frac{1}{2}} \text{excess kurtosis} = -\frac{6}{3+\nu}$$

For the unsymmetric cases ($\alpha \neq \beta$) the following limits (with only the noted variable approaching the limit) can be obtained from the above expressions:

$$\lim_{\alpha \to 0} \text{excess kurtosis} = \lim_{\beta \to 0} \text{excess kurtosis} = \lim_{\mu \to 0} \text{excess kurtosis} = \lim_{\mu \to 1} \text{excess kurtosis} = \infty$$

$$\lim_{\alpha \to \infty} \text{excess kurtosis} = \frac{6}{\beta}, \quad \lim_{\beta \to 0}(\lim_{\alpha \to \infty} \text{excess kurtosis}) = \infty, \quad \lim_{\beta \to \infty}(\lim_{\alpha \to \infty} \text{excess kurtosis}) = 0$$

$$\lim_{\beta \to \infty} \text{excess kurtosis} = \frac{6}{\alpha}, \quad \lim_{\alpha \to 0}(\lim_{\beta \to \infty} \text{excess kurtosis}) = \infty, \quad \lim_{\alpha \to \infty}(\lim_{\beta \to \infty} \text{excess kurtosis}) = 0$$

$$\lim_{\nu \to 0} \text{excess kurtosis} = -6 + \frac{1}{\mu(1-\mu)}, \quad \lim_{\mu \to 0}(\lim_{\nu \to 0} \text{excess kurtosis}) = \infty, \quad \lim_{\mu \to 1}(\lim_{\nu \to 0} \text{excess kurtosis}) = \infty$$

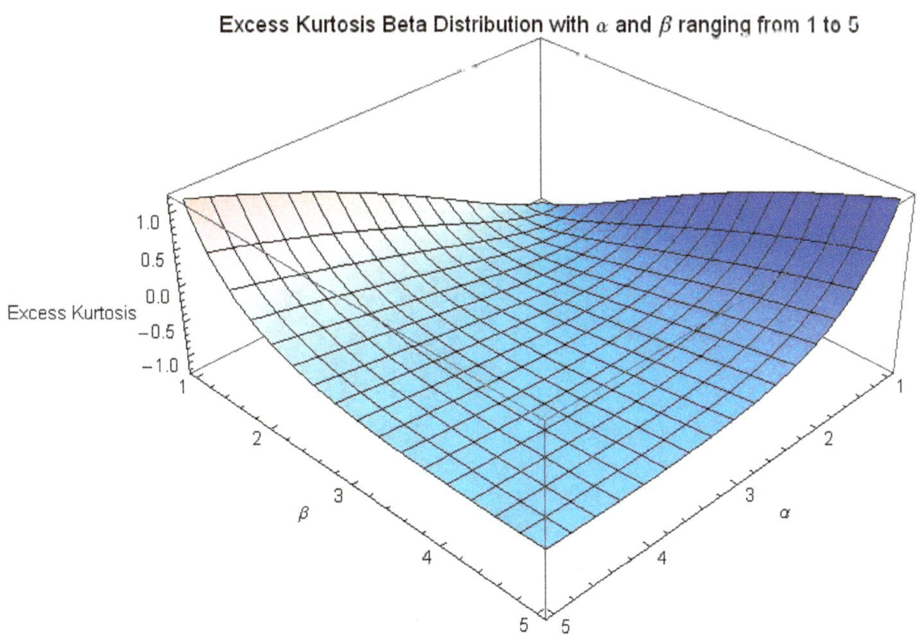

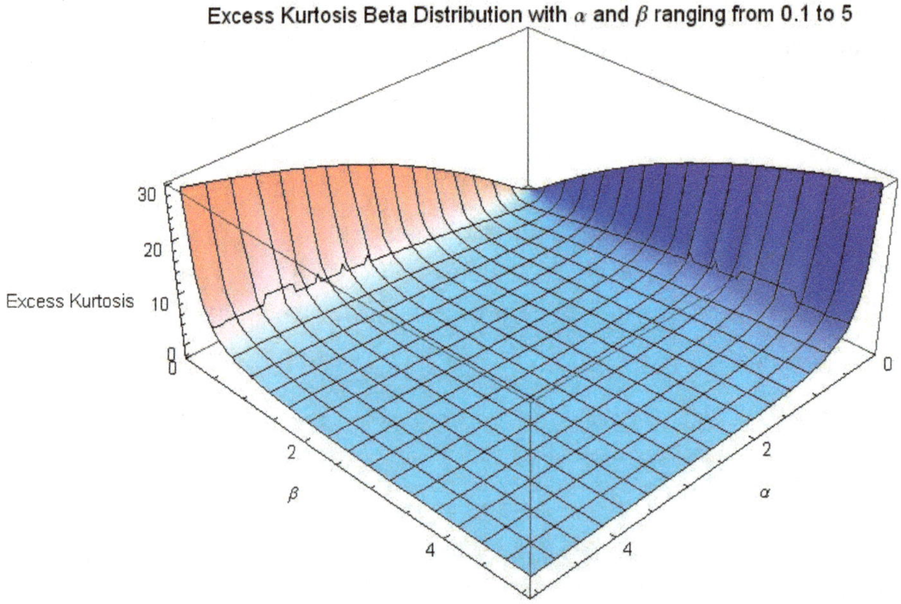

Characteristic Function

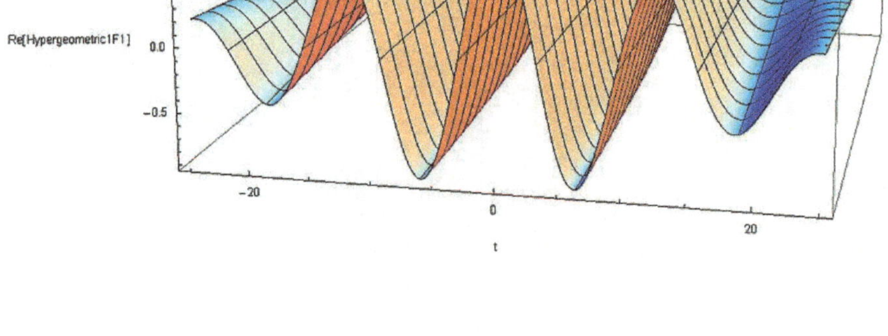

Re(characteristic function) symmetric case $\alpha = \beta$ ranging from 25 to 0

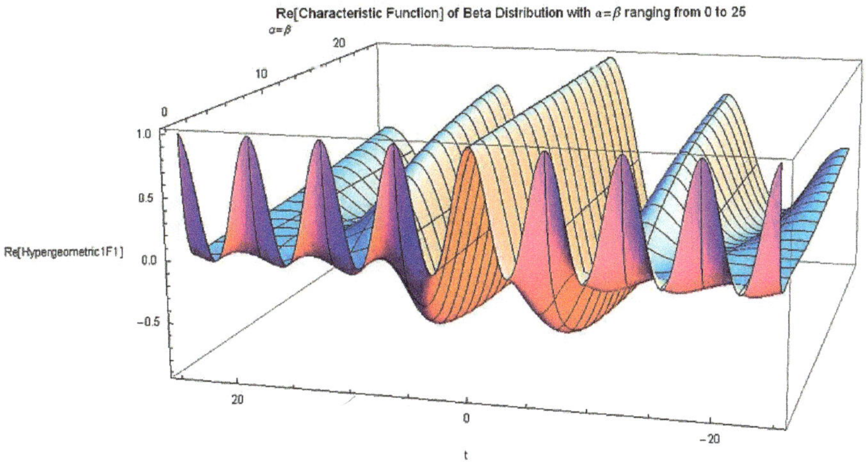

Re(characteristic function) symmetric case α = β ranging from 0 to 25.

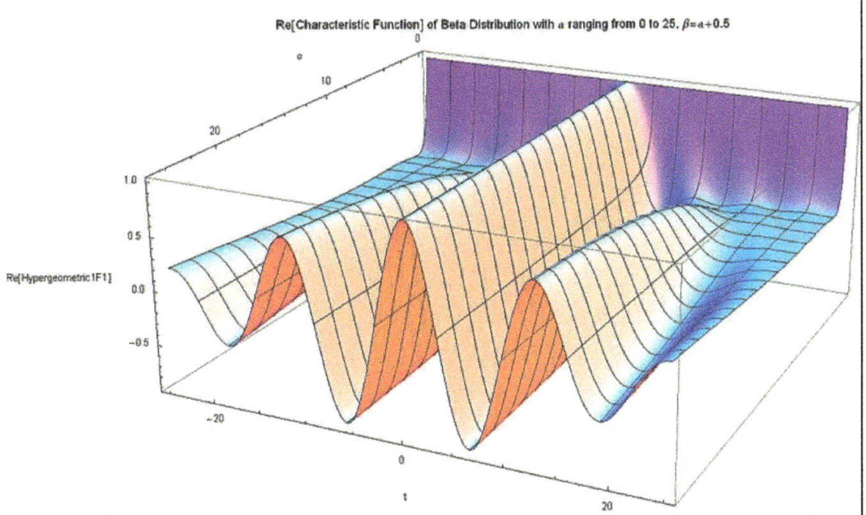

Re(characteristic function) β = α + 1/2; α ranging from 25 to 0.

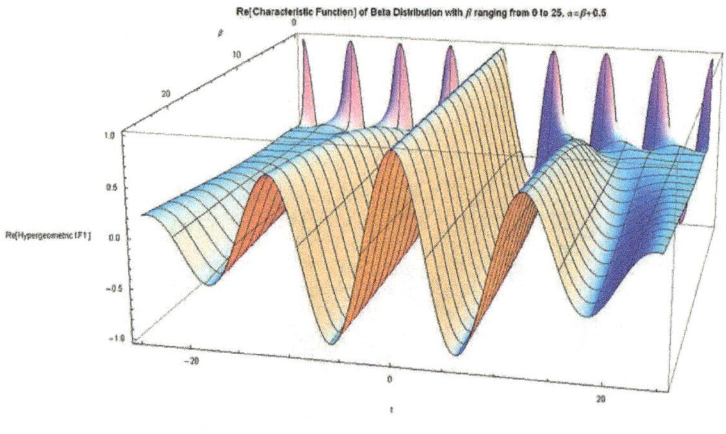

Re(characteristic function) α = β + 1/2; β ranging from 25 to 0

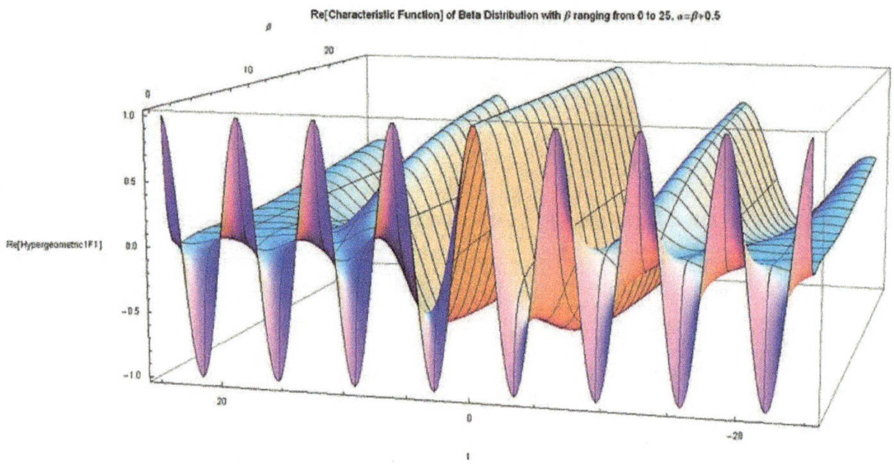

Re(characteristic function) α = β + 1/2; β ranging from 0 to 25.

The characteristic function is the Fourier transform of the probability density function. The characteristic function of the beta distribution is Kummer's confluent hypergeometric function (of the first kind):

$$\begin{aligned}
\varphi_X(\alpha;\beta;t) &= \mathrm{E}\left[e^{itX}\right] \\
&= \int_0^1 e^{itx} f(x;\alpha,\beta)\,dx \\
&= {}_1F_1(\alpha;\alpha+\beta;it) \\
&= \sum_{n=0}^{\infty} \frac{\alpha^{(n)}(it)^n}{(\alpha+\beta)^{(n)}n!} \\
&= 1+\sum_{k=1}^{\infty}\left(\prod_{r=0}^{k-1}\frac{\alpha+r}{\alpha+\beta+r}\right)\frac{(it)^k}{k!}
\end{aligned}$$

where

$x^{(n)} = x(x+1)(x+2)\cdots(x+n-1)$

is the rising factorial, also called the "Pochhammer symbol". The value of the characteristic function for t = 0, is one:

$$\varphi_X(\alpha;\beta;0) = {}_1F_1(\alpha;\alpha+\beta;0) = 1.$$

Also, the real and imaginary parts of the characteristic function enjoy the following symmetries with respect to the origin of variable t:

$$\mathrm{Re}\left[{}_1F_1(\alpha;\alpha+\beta;it)\right] = \mathrm{Re}\left[{}_1F_1(\alpha;\alpha+\beta;-it)\right]$$
$$\mathrm{Im}\left[{}_1F_1(\alpha;\alpha+\beta;it)\right] = -\mathrm{Im}\left[{}_1F_1(\alpha;\alpha+\beta;-it)\right]$$

The symmetric case α = β simplifies the characteristic function of the beta distribution to a Bessel function, since in the special case α + β = 2α the confluent hypergeometric function (of the first kind) reduces to a Bessel function (the modified Bessel function of the first kind $I_{\alpha-\frac{1}{2}}$ using Kummer's second transformation as follows:

$$\begin{aligned}{}_1F_1(\alpha;2\alpha;it) &= e^{\frac{it}{2}}\,{}_0F_1\left(;\alpha+\tfrac{1}{2};\frac{(it)^2}{16}\right)\\ &= e^{\frac{it}{2}}\left(\frac{it}{4}\right)^{\frac{1}{2}-\alpha}\Gamma\left(\alpha+\tfrac{1}{2}\right)I_{\alpha-\frac{1}{2}}\left(\frac{it}{2}\right)\end{aligned}$$

In the accompanying plots, the real part (Re) of the characteristic function of the beta distribution is displayed for symmetric (α = β) and skewed (α ≠ β) cases.

Other Moments

Moment Generating Function

It also follows that the moment generating function is

$$\begin{aligned}M_X(\alpha;\beta;t) &= \mathrm{E}\left[e^{tX}\right]\\ &= \int_0^1 e^{tx} f(x;\alpha,\beta)\,dx\\ &= {}_1F_1(\alpha;\alpha+\beta;t)\\ &= \sum_{n=0}^{\infty}\frac{\alpha^{(n)}}{(\alpha+\beta)^{(n)}}\frac{t^n}{n!}\\ &= 1+\sum_{k=1}^{\infty}\left(\prod_{r=0}^{k-1}\frac{\alpha+r}{\alpha+\beta+r}\right)\frac{t^k}{k!}\end{aligned}$$

In particular $M_X(\alpha;\beta;0) = 1$.

Higher Moments

Using the moment generating function, the k-th raw moment is given by the factor

$$\prod_{r=0}^{k-1} \frac{\alpha + r}{\alpha + \beta + r}$$

multiplying the (exponential series) term $\left(\frac{t^k}{k!}\right)$ in the series of the moment generating function

$$\mathrm{E}[X^k] = \frac{\alpha^{(k)}}{(\alpha+\beta)^{(k)}} = \prod_{r=0}^{k-1} \frac{\alpha + r}{\alpha + \beta + r}$$

where $(x)^{(k)}$ is a Pochhammer symbol representing rising factorial. It can also be written in a recursive form as

$$\mathrm{E}[X^k] = \frac{\alpha + k - 1}{\alpha + \beta + k - 1} \mathrm{E}[X^{k-1}].$$

Since the moment generating function $M_X(\alpha; \beta; \cdot)$ has a positive radius of convergence, the beta distribution is determined by its moments.

Moments of Transformed Random Variables

Moments of Linearly Transformed, Product and Inverted Random Variables

One can also show the following expectations for a transformed random variable, where the random variable X is Beta-distributed with parameters α and β: X ~ Beta(α, β). The expected value of the variable (1−X) is the mirror-symmetry of the expected value based on X:

$$\mathrm{E}[1 - X] = \frac{\beta}{\alpha + \beta}$$

$$\mathrm{E}[X(1 - X)] = \mathrm{E}[(1 - X)X] = \frac{\alpha\beta}{(\alpha+\beta)(\alpha+\beta+1)}$$

Due to the mirror-symmetry of the probability density function of the beta distribution, the variances based on variables X and (1−X) are identical, and the covariance on X(1-X) is the negative of the variance:

$$\mathrm{var}[(1 - X)] = \mathrm{var}[X] = -\operatorname{cov}[X, (1 - X)] = \frac{\alpha\beta}{(\alpha+\beta)^2(\alpha+\beta+1)}$$

These are the expected values for inverted variables, (these are related to the harmonic means, see section titled "Harmonic mean"):

$$\mathrm{E}\left[\frac{1}{X}\right] = \frac{\alpha+\beta-1}{\alpha-1} \text{ if } \alpha > 1$$

$$\mathrm{E}\left[\frac{1}{1-X}\right] = \frac{\alpha+\beta-1}{\beta-1} \text{ if } \beta > 1$$

The following transformation by dividing the variable X by its mirror-image X/(1−X) results in the expected value of the "inverted beta distribution" or beta prime distribution (also known as beta distribution of the second kind or Pearson›s Type VI):

$$\mathrm{E}\left[\frac{X}{1-X}\right] = \frac{\alpha}{\beta-1} \text{ if } \beta > 1$$
$$\mathrm{E}\left[\frac{1-X}{X}\right] = \frac{\beta}{\alpha-1} \text{ if } \alpha > 1$$

Variances of these transformed variables can be obtained by integration, as the expected values of the second moments centered on the corresponding variables:

$$\mathrm{var}\left[\frac{1}{X}\right] = \mathrm{E}\left[\left(\frac{1}{X} - \mathrm{E}\left[\frac{1}{X}\right]\right)^2\right] =$$
$$\mathrm{var}\left[\frac{1-X}{X}\right] = \mathrm{E}\left[\left(\frac{1-X}{X} - \mathrm{E}\left[\frac{1-X}{X}\right]\right)^2\right] = \frac{\beta(\alpha+\beta-1)}{(\alpha-2)(\alpha-1)^2} \text{ if } \alpha > 2$$

The following variance of the variable X divided by its mirror-image (X/(1−X) results in the variance of the "inverted beta distribution" or beta prime distribution (also known as beta distribution of the second kind or Pearson›s Type VI):

$$\mathrm{var}\left[\frac{1}{1-X}\right] = \mathrm{E}\left[\left(\frac{1}{1-X} - \mathrm{E}\left[\frac{1}{1-X}\right]\right)^2\right] = \mathrm{var}\left[\frac{X}{1-X}\right] =$$
$$\mathrm{E}\left[\left(\frac{X}{1-X} - \mathrm{E}\left[\frac{X}{1-X}\right]\right)^2\right] = \frac{\alpha(\alpha+\beta-1)}{(\beta-2)(\beta-1)^2} \text{ if } \beta > 2$$

The covariances are:

$$\mathrm{cov}\left[\frac{1}{X}, \frac{1}{1-X}\right] = \mathrm{cov}\left[\frac{1-X}{X}, \frac{X}{1-X}\right] = \mathrm{cov}\left[\frac{1}{X}, \frac{X}{1-X}\right] = \mathrm{cov}\left[\frac{1-X}{X}, \frac{1}{1-X}\right] = \frac{\alpha+\beta-1}{(\alpha-1)(\beta-1)} \text{ if } \alpha, \beta > 1$$

These expectations and variances appear in the four-parameter Fisher information matrix (section titled "Fisher information," "four parameters").

Moments of Logarithmically Transformed Random Variables

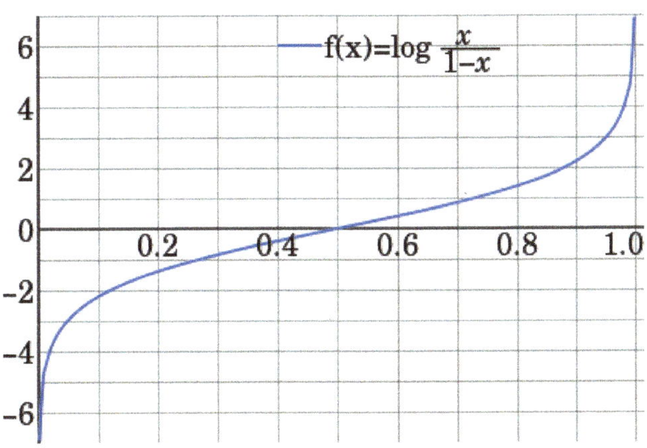

Plot of logit(X) = ln(X/(1−X)) (vertical axis) vs. X in the domain of 0 to 1 (horizontal axis). Logit transformations are interesting, as they usually transform various shapes (including J-shapes) into (usually skewed) bell-shaped densities over the logit variable, and they may remove the end singularities over the original variable.

Expected values for logarithmic transformations (useful for maximum likelihood estimates, see section titled "Parameter estimation, Maximum likelihood" below) are discussed in this section. The following logarithmic linear transformations are related to the geometric means G_X and $G_{(1-X)}$ (see section titled "Geometric mean"):

$$\mathrm{E}[\ln(X)] = \psi(\alpha) - \psi(\alpha+\beta) = -\mathrm{E}\left[\ln\left(\frac{1}{X}\right)\right],$$

$$\mathrm{E}[\ln(1-X)] = \psi(\beta) - \psi(\alpha+\beta) = -\mathrm{E}\left[\ln\left(\frac{1}{1-X}\right)\right].$$

Where the **digamma function** $\psi(\alpha)$ is defined as the logarithmic derivative of the gamma function:

$$\psi(\alpha) = \frac{d\ln\Gamma(\alpha)}{d\alpha}$$

Logit transformations are interesting, as they usually transform various shapes (including J-shapes) into (usually skewed) bell-shaped densities over the logit variable, and they may remove the end singularities over the original variable:

$$\mathrm{E}\left[\ln\left(\frac{X}{1-X}\right)\right] = \psi(\alpha) - \psi(\beta) = \mathrm{E}[\ln(X)] + \mathrm{E}\left[\ln\left(\frac{1}{1-X}\right)\right],$$

$$\mathrm{E}\left[\ln\left(\frac{1-X}{X}\right)\right] = \psi(\beta) - \psi(\alpha) = -\mathrm{E}\left[\ln\left(\frac{X}{1-X}\right)\right].$$

Johnson considered the distribution of the logit - transformed variable ln(X/1−X), including its moment generating function and approximations for large values of the shape parameters. This transformation extends the finite support [0, 1] based on the original variable X to infinite support in both directions of the real line $(-\infty, +\infty)$.

Higher order logarithmic moments can be derived by using the representation of a beta distribution as a proportion of two Gamma distributions and differentiating through the integral. They can be expressed in terms of higher order poly-gamma functions as follows:

$$\begin{aligned}\mathrm{E}\left[\ln^2(X)\right] &= (\psi(\alpha) - \psi(\alpha+\beta))^2 + \psi_1(\alpha) - \psi_1(\alpha+\beta),\\ \mathrm{E}\left[\ln^2(1-X)\right] &= (\psi(\beta) - \psi(\alpha+\beta))^2 + \psi_1(\beta) - \psi_1(\alpha+\beta),\\ \mathrm{E}[\ln(X)\ln(1-X)] &= (\psi(\alpha) - \psi(\alpha+\beta))(\psi(\beta) - \psi(\alpha+\beta)) - \psi_1(\alpha+\beta).\end{aligned}$$

therefore the variance of the logarithmic variables and covariance of ln(X) and ln(1−X) are:

$$\mathrm{cov}[\ln(X), \ln(1-X)] = \mathrm{E}[\ln(X)\ln(1-X)] - \mathrm{E}[\ln(X)]\,\mathrm{E}[\ln(1-X)] = -\psi_1(\alpha+\beta)$$

$$\begin{aligned}\mathrm{var}[\ln X] &= \mathrm{E}[\ln^2(X)] - (\mathrm{E}[\ln(X)])^2\\ &= \psi_1(\alpha) - \psi_1(\alpha+\beta)\\ &= \psi_1(\alpha) + \mathrm{cov}[\ln(X), \ln(1-X)]\end{aligned}$$

$$\begin{aligned}\mathrm{var}[\ln(1-X)] &= \mathrm{E}[\ln^2(1-X)] - (\mathrm{E}[\ln(1-X)])^2\\ &= \psi_1(\beta) - \psi_1(\alpha+\beta)\\ &= \psi_1(\beta) + \mathrm{cov}[\ln(X), \ln(1-X)]\end{aligned}$$

where the **trigamma function**, denoted $\psi_1(\alpha)$, is the second of the polygamma functions, and is defined as the derivative of the digamma function:

$$\psi_1(\alpha) = \frac{d^2 \ln \Gamma(\alpha)}{d\alpha^2} = \frac{d\psi(\alpha)}{d\alpha}.$$

The variances and covariance of the logarithmically transformed variables X and (1−X) are different, in general, because the logarithmic transformation destroys the mirror-symmetry of the original variables X and (1−X), as the logarithm approaches negative infinity for the variable approaching zero.

These logarithmic variances and covariance are the elements of the Fisher information matrix for the beta distribution. They are also a measure of the curvature of the log likelihood function (see section on Maximum likelihood estimation).

The variances of the log inverse variables are identical to the variances of the log variables:

$$\begin{aligned}\operatorname{var}\left[\ln\left(\frac{1}{X}\right)\right] &= \operatorname{var}[\ln(X)] = \psi_1(\alpha) - \psi_1(\alpha+\beta),\\ \operatorname{var}\left[\ln\left(\frac{1}{1-X}\right)\right] &= \operatorname{var}[\ln(1-X)] = \psi_1(\beta) - \psi_1(\alpha+\beta),\\ \operatorname{cov}\left[\ln\left(\frac{1}{X}\right), \ln\left(\frac{1}{1-X}\right)\right] &= \operatorname{cov}[\ln(X), \ln(1-X)] = -\psi_1(\alpha+\beta).\end{aligned}$$

It also follows that the variances of the logit transformed variables are:

$$\operatorname{var}\left[\ln\left(\frac{X}{1-X}\right)\right] = \operatorname{var}\left[\ln\left(\frac{1-X}{X}\right)\right] = -\operatorname{cov}\left[\ln\left(\frac{X}{1-X}\right), \ln\left(\frac{1-X}{X}\right)\right] = \psi_1(\alpha) + \psi_1(\beta)$$

Quantities of Informaation (Entropy)

Given a beta distributed random variable, X ~ Beta(α, β), the differential entropy of X is(measured in nats), the expected value of the negative of the logarithm of the probability density function:

$$\begin{aligned}h(X) &= \mathrm{E}[-\ln(f(x;\alpha,\beta))]\\ &= \int_0^1 -f(x;\alpha,\beta)\ln(f(x;\alpha,\beta))dx\\ &= \ln(\mathrm{B}(\alpha,\beta)) - (\alpha-1)\psi(\alpha) - (\beta-1)\psi(\beta) + (\alpha+\beta-2)\psi(\alpha+\beta)\end{aligned}$$

where f(x; α, β) is the probability density function of the beta distribution:

$$f(x;\alpha,\beta) = \frac{1}{\mathrm{B}(\alpha,\beta)} x^{\alpha-1}(1-x)^{\beta-1}$$

The digamma function ψ appears in the formula for the differential entropy as a consequence of Euler's integral formula for the harmonic numbers which follows from the integral:

$$\int_0^1 \frac{1-x^{\alpha-1}}{1-x}dx = \psi(\alpha) - \psi(1)$$

The differential entropy of the beta distribution is negative for all values of α and β greater than zero, except at α = β = 1 (for which values the beta distribution is the same as the uniform distribution), where the differential entropy reaches its maximum value of zero. It is to be expected that the maximum entropy should take place when the beta distribution becomes equal to the uniform distribution, since uncertainty is maximal when all possible events are equiprobable.

For α or β approaching zero, the differential entropy approaches its minimum value of negative infinity. For (either or both) α or β approaching zero, there is a maximum amount of order: all the probability density is concentrated at the ends, and there is zero probability density at points located between the ends. Similarly for (either or both) α or β approaching infinity, the differential entropy approaches its minimum value of negative infinity, and a maximum amount of order. If either α or β approaches infinity (and the other is finite) all the probability density is concentrated at an end, and the probability density is zero everywhere else. If both shape parameters are equal (the symmetric case), α = β, and they approach infinity simultaneously, the probability density becomes a spike (Dirac delta function) concentrated at the middle x = 1/2, and hence there is 100% probability at the middle x = 1/2 and zero probability everywhere else.

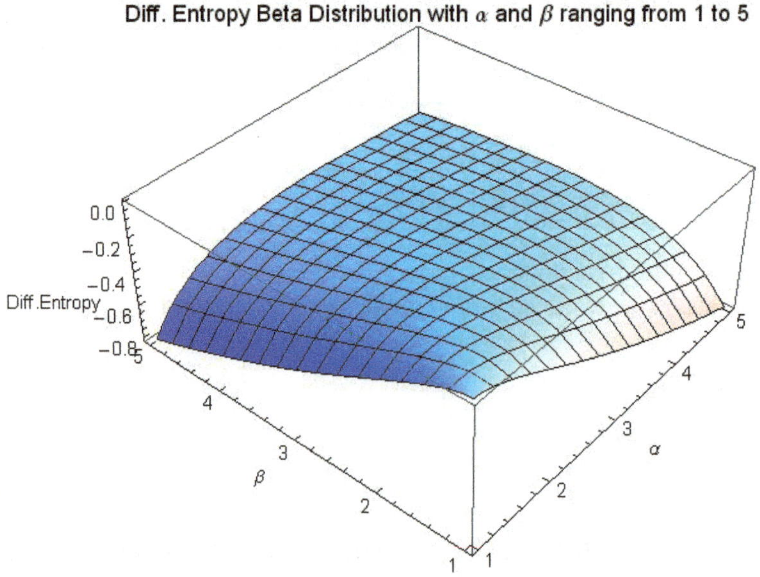

Diff. Entropy Beta Distribution with α and β ranging from 0.1 to 5

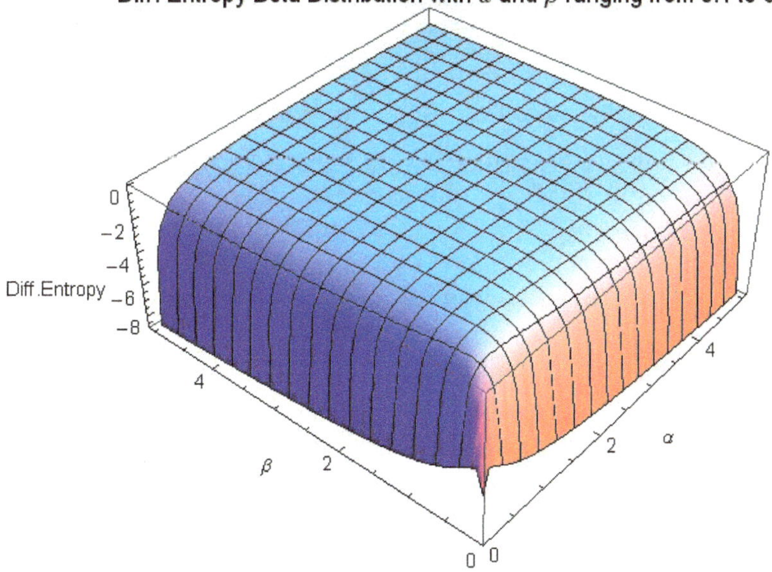

The (continuous case) differential entropy was introduced by Shannon in his original paper (where he named it the "entropy of a continuous distribution"), as the concluding part of the same paper where he defined the discrete entropy. It is known since then that the differential entropy may differ from the infinitesimal limit of the discrete entropy by an infinite offset, therefore the differential entropy can be negative (as it is for the beta distribution). What really matters is the relative value of entropy.

Given two beta distributed random variables, $X_1 \sim \text{Beta}(\alpha, \beta)$ and $X_2 \sim \text{Beta}(\alpha', \beta')$, the cross entropy is (measured in nats)

$$\begin{aligned} H(X_1, X_2) &= \int_0^1 -f(x;\alpha,\beta)\ln(f(x;\alpha',\beta'))dx \\ &= \ln(\mathrm{B}(\alpha',\beta')) - (\alpha'-1)\psi(\alpha) - (\beta'-1)\psi(\beta) + (\alpha'+\beta'-2)\psi(\alpha+\beta). \end{aligned}$$

The cross entropy has been used as an error metric to measure the distance between two hypotheses. Its absolute value is minimum when the two distributions are identical. It is the information measure most closely related to the log maximum likelihood (see section on "Parameter estimation. Maximum likelihood estimation")).

The relative entropy, or Kullback–Leibler divergence $D_{KL}(X_1, X_2)$, is a measure of the inefficiency of assuming that the distribution is $X_2 \sim \text{Beta}(\alpha', \beta')$ when the distribution is really $X_1 \sim \text{Beta}(\alpha, \beta)$. It is defined as follows (measured in nats).

$$\begin{aligned} D_{\mathrm{KL}}(X_1, X_2) &= \int_0^1 f(x;\alpha,\beta)\ln\left(\frac{f(x;\alpha,\beta)}{f(x;\alpha',\beta')}\right)dx \\ &= \left(\int_0^1 f(x;\alpha,\beta)\ln(f(x;\alpha,\beta))dx\right) - \left(\int_0^1 f(x;\alpha,\beta)\ln(f(x;\alpha',\beta'))dx\right) \\ &= -h(X_1) + H(X_1, X_2) \\ &= \ln\left(\frac{\mathrm{B}(\alpha',\beta')}{\mathrm{B}(\alpha,\beta)}\right) + (\alpha-\alpha')\psi(\alpha) + (\beta-\beta')\psi(\beta) + (\alpha'-\alpha+\beta'-\beta)\psi(\alpha+\beta). \end{aligned}$$

The relative entropy, or Kullback–Leibler divergence, is always non-negative. A few numerical examples follow:

- $X_1 \sim$ Beta(1, 1) and $X_2 \sim$ Beta(3, 3); $D_{KL}(X_1, X_2) = 0.598803$; $D_{KL}(X_2, X_1) = 0.267864$; $h(X_1) = 0$; $h(X_2) = -0.267864$
- $X_1 \sim$ Beta(3, 0.5) and $X_1 \sim$ Beta(0.5, 3); $D_{KL}(X_1, X_2) = 7.21574$; $D_{KL}(X_2, X_1) = 7.21574$; $h(X_1) = -1.10805$; $h(X_2) = -1.10805$.

The Kullback–Leibler divergence is not symmetric $D_{KL}(X_1, X_2) \neq D_{KL}(X_2, X_1)$ for the case in which the individual beta distributions Beta(1, 1) and Beta(3, 3) are symmetric, but have different entropies $h(X_1) \neq h(X_2)$. The value of the Kullback divergence depends on the direction traveled: whether going from a higher (differential) entropy to a lower (differential) entropy or the other way around. In the numerical example above, the Kullback divergence measures the inefficiency of assuming that the distribution is (bell-shaped) Beta(3, 3), rather than (uniform) Beta(1, 1). The "h" entropy of Beta(1, 1) is higher than the "h" entropy of Beta(3, 3) because the uniform distribution Beta(1, 1) has a maximum amount of disorder. The Kullback divergence is more than two times higher (0.598803 instead of 0.267864) when measured in the direction of decreasing entropy: the direction that assumes that the (uniform) Beta(1, 1) distribution is (bell-shaped) Beta(3, 3) rather than the other way around. In this restricted sense, the Kullback divergence is consistent with the second law of thermodynamics.

The Kullback–Leibler divergence is symmetric $D_{KL}(X_1, X_2) = D_{KL}(X_2, X_1)$ for the skewed cases Beta(3, 0.5) and Beta(0.5, 3) that have equal differential entropy $h(X_1) = h(X_2)$.

The symmetry condition:

$$D_{KL}(X_1, X_2) = D_{KL}(X_2, X_1), \text{ if } h(X_1) = h(X_2), \text{ for (skewed) } \alpha \neq \beta$$

follows from the above definitions and the mirror-symmetry $f(x; \alpha, \beta) = f(1-x; \alpha, \beta)$ enjoyed by the beta distribution.

Relationships between Statistical Measures

Mean, mode and Median Relationship

If $1 < \alpha < \beta$ then mode $\leq$ median $\leq$ mean. Expressing the mode (only for $\alpha, \beta > 1$), and the mean in terms of α and β:

$$\frac{\alpha - 1}{\alpha + \beta - 2} \leq \text{median} \leq \frac{\alpha}{\alpha + \beta},$$

If $1 < \beta < \alpha$ then the order of the inequalities are reversed. For $\alpha, \beta > 1$ the absolute distance between the mean and the median is less than 5% of the distance between the maximum and minimum values of x. On the other hand, the absolute distance between the mean and the mode can reach 50% of the distance between the maximum and minimum values of x, for the (pathological) case of $\alpha = 1$ and $\beta = 1$ (for which values the beta distribution approaches the uniform distribution and the differential entropy approaches its maximum value, and hence maximum "disorder").

For example, for $\alpha = 1.0001$ and $\beta = 1.00000001$:

- mode = 0.9999; PDF(mode) = 1.00010
- mean = 0.500025; PDF(mean) = 1.00003

- median = 0.500035; PDF(median) = 1.00003
- mean − mode = −0.499875
- mean − median = -9.65538×10^{-6}

(where PDF stands for the value of the probability density function)

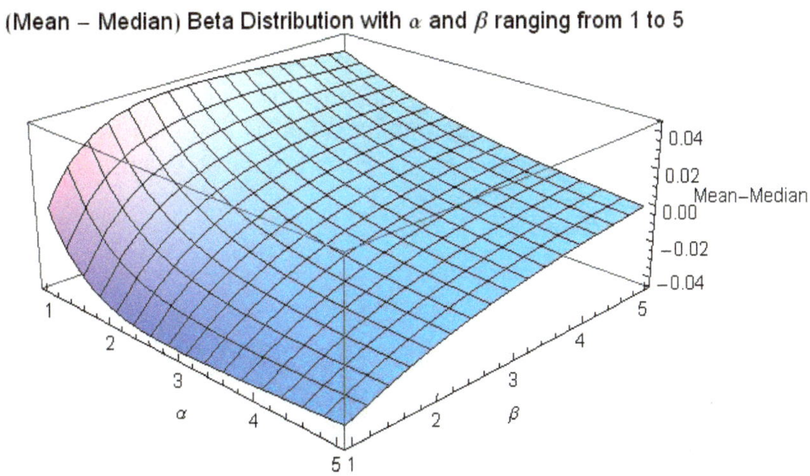

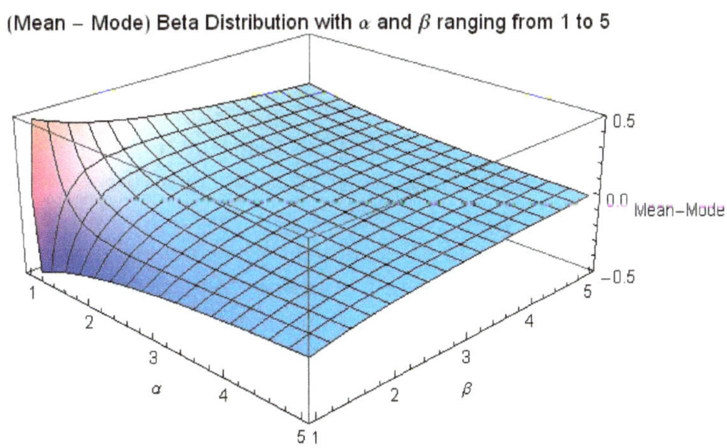

Mean, Geometric mean and Harmonic Mean Relationship

:Mean, Median, Geometric Mean and Harmonic Mean for Beta distribution with $0 < \alpha = \beta < 5$

It is known from the inequality of arithmetic and geometric means that the geometric mean is lower than the mean. Similarly, the harmonic mean is lower than the geometric mean. The accompanying plot shows that for $\alpha = \beta$, both the mean and the median are exactly equal to 1/2, regardless of the value of $\alpha = \beta$, and the mode is also equal to 1/2 for $\alpha = \beta > 1$, however the geometric and harmonic means are lower than 1/2 and they only approach this value asymptotically as $\alpha = \beta \to \infty$.

Kurtosis Bounded by the Square of the Skewness

Beta distribution α and β parameters vs. excess Kurtosis and squared Skewness

As remarked by Feller, in the Pearson system the beta probability density appears as type I (any difference between the beta distribution and Pearson's type I distribution is only superficial and it makes no difference for the following discussion regarding the relationship between kurtosis and skewness). Karl Pearson showed, in Plate 1 of his paper published in 1916, a graph with the kurtosis as the vertical axis (ordinate) and the square of the skewness as the horizontal axis (abscissa), in which a number of distributions were displayed. The region occupied by the beta distribution is bounded by the following two lines in the (skewness^2,kurtosis) plane, or the (skewness^2,excess kurtosis) plane:

$$(\text{skewness})^2 + 1 < \text{kurtosis} < \frac{3}{2}(\text{skewness})^2 + 3$$

or, equivalently,

$$(\text{skewness})^2 - 2 < \text{excess kurtosis} < \frac{3}{2}(\text{skewness})^2$$

(At a time when there were no powerful digital computers), Karl Pearson accurately computed further boundaries, for example, separating the "U-shaped" from the "J-shaped" distributions. The lower boundary line (excess kurtosis + 2 − skewness^2 = 0) is produced by skewed "U-shaped" beta distributions with both values of shape parameters α and β close to zero. The upper boundary line (excess kurtosis − (3/2) skewness^2 = 0) is produced by extremely skewed distributions with very large values of one of the parameters and very small values of the other parameter. Karl Pearson showed that this upper boundary line (excess kurtosis − (3/2) skewness^2 = 0) is also the intersection with Pearson's distribution III, which has unlimited support in one direction (towards positive infinity), and can be bell-shaped or J-shaped. His son, Egon Pearson, showed that the region (in the kurtosis/squared-skewness plane) occupied by the beta distribution (equivalently, Pearson›s distribution I) as it approaches this boundary (excess kurtosis − (3/2) skewness^2 = 0) is shared with the noncentral chi-squared distribution. Karl Pearson (Pearson 1895, pp. 357, 360, 373–376) also showed that the gamma distribution is a Pearson type III distribution. Hence this boundary line for Pearson›s type III distribution is known as the gamma line. (This can be shown from the fact that the excess kurtosis of the gamma distribution is 6/k and the square of the skewness is 4/k, hence (excess kurtosis − (3/2) skewness^2 = 0) is identically satisfied by the gamma distribution regardless of the value of the parameter "k"). Pearson later noted that the chi-squared distribution is a special case of Pearson's type III and also shares this boundary line (as it is apparent from the fact that for the chi-squared distribution the excess kurtosis is 12/k and the square of the skewness is 8/k, hence (excess kurtosis − (3/2) skewness^2 = 0) is identically satisfied regardless of the value of the parameter "k"). This is to be expected, since the chi-squared distribution $X \sim \chi^2(k)$ is a special case of the gamma distribution, with parametrization $X \sim \Gamma(k/2, 1/2)$ where k is a positive integer that specifies the "number of degrees of freedom" of the chi-squared distribution.

An example of a beta distribution near the upper boundary (excess kurtosis − (3/2) skewness^2 = 0) is given by α = 0.1, β = 1000, for which the ratio (excess kurtosis)/

(skewness^2) = 1.49835 approaches the upper limit of 1.5 from below. An example of a beta distribution near the lower boundary (excess kurtosis + 2 − skewness^2 = 0) is given by α= 0.0001, β = 0.1, for which values the expression (excess kurtosis + 2)/(skewness^2) = 1.01621 approaches the lower limit of 1 from above. In the infinitesimal limit for both α and β approaching zero symmetrically, the excess kurtosis reaches its minimum value at −2. This minimum value occurs at the point at which the lower boundary line intersects the vertical axis (ordinate). (However, in Pearson's original chart, the ordinate is kurtosis, instead of excess kurtosis, and it increases downwards rather than upwards).

Values for the skewness and excess kurtosis below the lower boundary (excess kurtosis + 2 − skewness^2 = 0) cannot occur for any distribution, and hence Karl Pearson appropriately called the region below this boundary the "impossible region." The boundary for this "impossible region" is determined by (symmetric or skewed) bimodal "U"-shaped distributions for which parameters α and β approach zero and hence all the probability density is concentrated at the ends: x = 0, 1 with practically nothing in between them. Since for α ≈ β ≈ 0 the probability density is concentrated at the two ends x = 0 and x = 1, this "impossible boundary" is determined by a 2-point distribution: the probability can only take 2 values (Bernoulli distribution), one value with probability p and the other with probability q = 1−p. For cases approaching this limit boundary with symmetry α = β, skewness ≈ 0, excess kurtosis ≈ −2 (this is the lowest excess kurtosis possible for any distribution), and the probabilities are p ≈ q ≈ 1/2. For cases approaching this limit boundary with skewness, excess kurtosis ≈ −2 + skewness^2, and the probability density is concentrated more at one end than the other end (with practically nothing in between), with probabilities $p = \frac{\beta}{\alpha+\beta}$ at the left end x = 0 and $q = 1 - p = \frac{\alpha}{\alpha+\beta}$ at the right end x = 1.

Symmetry

All statements are conditional on α, β > 0

- **Probability density function** reflection symmetry

  $$f(x;\alpha,\beta) = f(1-x;\beta,\alpha)$$

- **Cumulative distribution function** reflection symmetry plus unitary translation

  $$F(x;\alpha,\beta) = I_x(\alpha,\beta) = 1 - F(1-x;\beta,\alpha) = 1 - I_{1-x}(\beta,\alpha)$$

- **Mode** reflection symmetry plus unitary translation

  $$\text{mode}(\text{B}(\alpha,\beta)) = 1 - \text{mode}(\text{B}(\beta,\alpha)), \text{ if } \text{B}(\beta,\alpha) \neq \text{B}(1,1)$$

- **Median** reflection symmetry plus unitary translation

  $$\text{median}(\text{B}(\alpha,\beta)) = 1 - \text{median}(\text{B}(\beta,\alpha))$$

- **Mean** reflection symmetry plus unitary translation

  $$\mu(\text{B}(\alpha,\beta)) = 1 - \mu(\text{B}(\beta,\alpha))$$

- **Geometric Means** each is individually asymmetric, the following symmetry applies between the geometric mean based on *X* and the geometric mean based on its reflection (1-X)

  $$G_X(\text{B}(\alpha,\beta)) = G_{(1-X)}(\text{B}(\beta,\alpha))$$

- **Harmonic means each is individually asymmetric, the following symmetry applies between the harmonic mean based on** X **and the harmonic mean based on its** reflection **(1-X)**

$H_X(\mathrm{B}(\alpha,\beta)) = H_{(1-X)}(\mathrm{B}(\beta,\alpha))$ if $\alpha, \beta > 1$.

- Variance symmetry

$\mathrm{var}(\mathrm{B}(\alpha,\beta)) = \mathrm{var}(\mathrm{B}(\beta,\alpha))$

- **Geometric variances each is individually asymmetric, the following symmetry applies between the log geometric variance based on X and the log geometric variance based on its** reflection **(1-X)**

$\ln(\mathrm{var}_{GX}(\mathrm{B}(\alpha,\beta))) = \ln(\mathrm{var}_{G(1-X)}(\mathrm{B}(\beta,\alpha)))$

- Geometric covariance symmetry

$\ln \mathrm{cov}_{GX,(1-X)}(\mathrm{B}(\alpha,\beta)) = \ln \mathrm{cov}_{GX,(1-X)}(\mathrm{B}(\beta,\alpha))$

- Mean absolute deviation around the mean symmetry

$$\mathrm{E}[|X - E[X]|](\mathrm{B}(\alpha,\beta)) = \mathrm{E}[|X - E[X]|](\mathrm{B}(\beta,\alpha))$$

- **Skewness** skew-symmetry

$$\mathrm{skewness}(\mathrm{B}(\alpha,\beta)) = -\mathrm{skewness}(\mathrm{B}(\beta,\alpha))$$

- **Excess kurtosis** symmetry

$$\mathrm{excess\ kurtosis}(\mathrm{B}(\alpha,\beta)) = \mathrm{excess\ kurtosis}(\mathrm{B}(\beta,\alpha))$$

- **Characteristic function** symmetry of Real part (with respect to the origin of variable "t")

$$\mathrm{Re}[{}_1F_1(\alpha;\alpha+\beta;it)] = \mathrm{Re}[{}_1F_1(\alpha;\alpha+\beta;-it)]$$

- **Characteristic function** skew-symmetry of Imaginary part (with respect to the origin of variable "t")

$$\mathrm{Im}[{}_1F_1(\alpha;\alpha+\beta;it)] = -\mathrm{Im}[{}_1F_1(\alpha;\alpha+\beta;-it)]$$

- **Characteristic function** symmetry of Absolute value (with respect to the origin of variable "t")

$$\mathrm{Abs}[{}_1F_1(\alpha;\alpha+\beta;it)] = \mathrm{Abs}[{}_1F_1(\alpha;\alpha+\beta;-it)]$$

- **Differential entropy** symmetry

$$h(\mathrm{B}(\alpha,\beta)) = h(\mathrm{B}(\beta,\alpha))$$

- **Relative Entropy (also called Kullback–Leibler divergence)** symmetry

$$D_{\mathrm{KL}}(X_1,X_2) = D_{\mathrm{KL}}(X_2,X_1), \text{ if } h(X_1) = h(X_2), \text{ for (skewed) } \alpha \neq \beta$$

- **Fisher information matrix** symmetry

$$\mathcal{I}_{i,j} = \mathcal{I}_{j,i}$$

Geometry of the Probability Density Function

Inflection Points

Inflection point location versus α and β showing regions with one inflection point

Inflection point location versus α and β showing region with two inflection points

For certain values of the shape parameters α and β, the probability density function has inflection points, at which the curvature changes sign. The position of these inflection points can be useful as a measure of the dispersion or spread of the distribution.

Defining the following quantity:

$$\kappa = \frac{\sqrt{\frac{(\alpha-1)(\beta-1)}{\alpha+\beta-3}}}{\alpha+\beta-2}$$

Points of inflection occur, depending on the value of the shape parameters α and β, as follows:

- (α > 2, β > 2) The distribution is bell-shaped (symmetric for α = β and skewed otherwise), with **two inflection points**, equidistant from the mode:

$$x = \text{mode} \pm \kappa = \frac{\alpha - 1 \pm \sqrt{\frac{(\alpha-1)(\beta-1)}{\alpha+\beta-3}}}{\alpha+\beta-2}$$

- (α = 2, β > 2) The distribution is unimodal, positively skewed, right-tailed, with **one inflection point**, located to the right of the mode:

$$x = \text{mode} + \kappa = \frac{2}{\beta}$$

- (α > 2, β = 2) The distribution is unimodal, negatively skewed, left-tailed, with **one inflection point**, located to the left of the mode:

$$x = \text{mode} - \kappa = 1 - \frac{2}{\alpha}$$

- (1 < α < 2, β > 2, α+β>2) The distribution is unimodal, positively skewed, right-tailed, with **one inflection point**, located to the right of the mode:

$$x = \text{mode} + \kappa = \frac{\alpha - 1 + \sqrt{\frac{(\alpha-1)(\beta-1)}{\alpha+\beta-3}}}{\alpha+\beta-2}$$

- (0 < α < 1, 1 < β < 2) The distribution has a mode at the left end x = 0 and it is positively skewed, right-tailed. There is **one inflection point**, located to the right of the mode:

$$x = \frac{\alpha - 1 + \sqrt{\frac{(\alpha-1)(\beta-1)}{\alpha+\beta-3}}}{\alpha+\beta-2}$$

- (α > 2, 1 < β < 2) The distribution is unimodal negatively skewed, left-tailed, with **one inflection point**, located to the left of the mode:

$$x = \text{mode} - \kappa = \frac{\alpha - 1 - \sqrt{\frac{(\alpha-1)(\beta-1)}{\alpha+\beta-3}}}{\alpha + \beta - 2}$$

- (1 < α < 2, 0 < β < 1) The distribution has a mode at the right end x=1 and it is negatively skewed, left-tailed. There is **one inflection point**, located to the left of the mode:

$$x = \frac{\alpha - 1 - \sqrt{\frac{(\alpha-1)(\beta-1)}{\alpha+\beta-3}}}{\alpha + \beta - 2}$$

There are no inflection points in the remaining (symmetric and skewed) regions: U-shaped: (α, β < 1) upside-down-U-shaped: (1 < α < 2, 1 < β < 2), reverse-J-shaped (α < 1, β > 2) or J-shaped: (α > 2, β < 1)

The accompanying plots show the inflection point locations (shown vertically, ranging from 0 to 1) versus α and β (the horizontal axes ranging from 0 to 5). There are large cuts at surfaces intersecting the lines α = 1, β = 1, α = 2, and β = 2 because at these values the beta distribution change from 2 modes, to 1 mode to no mode.

Shapes

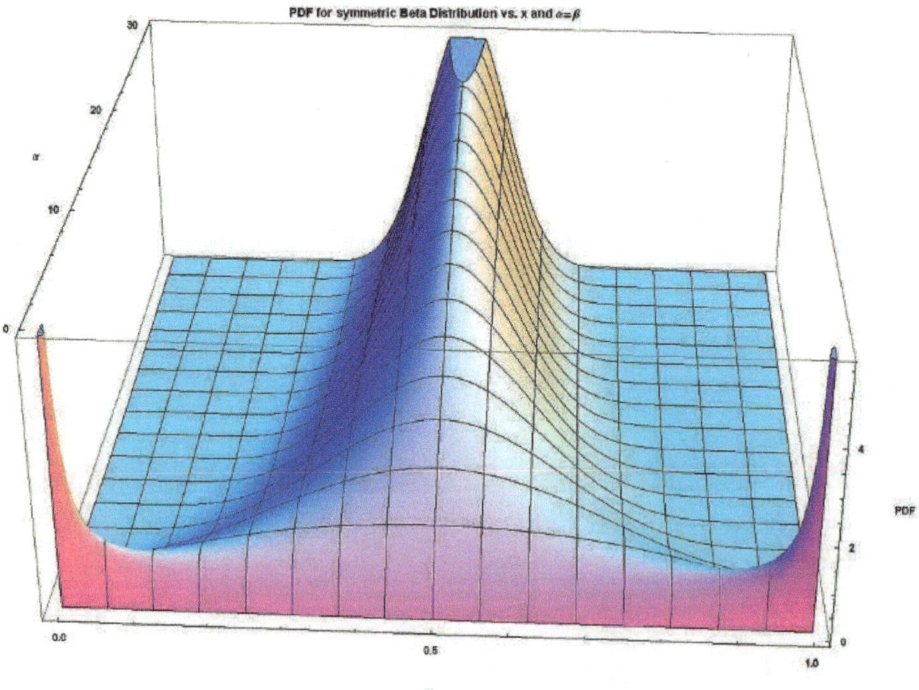

PDF for symmetric beta distribution vs. x and alpha=beta from 0 to 30

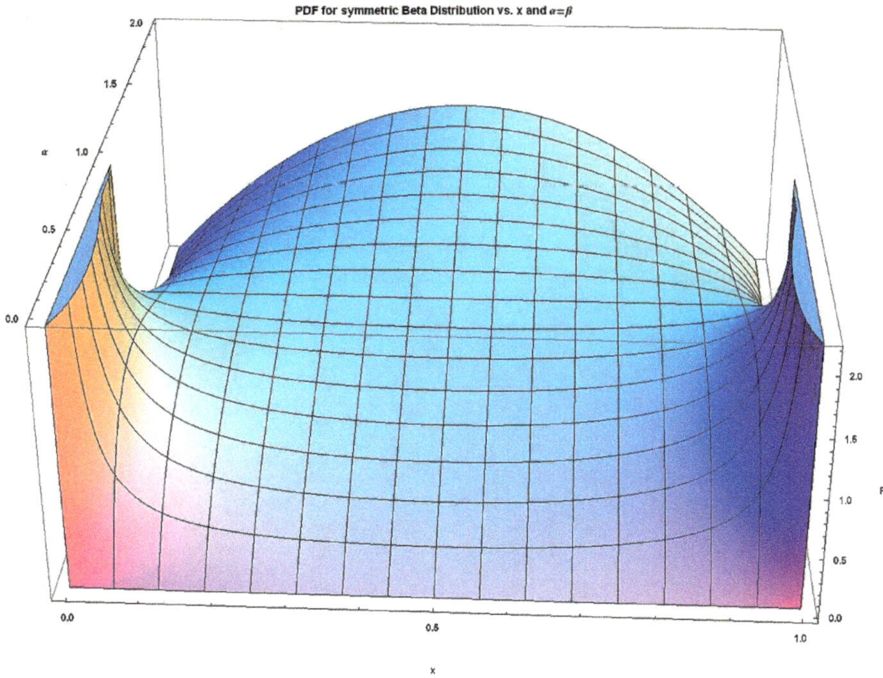

PDF for symmetric beta distribution vs. x and alpha=beta from 0 to 2

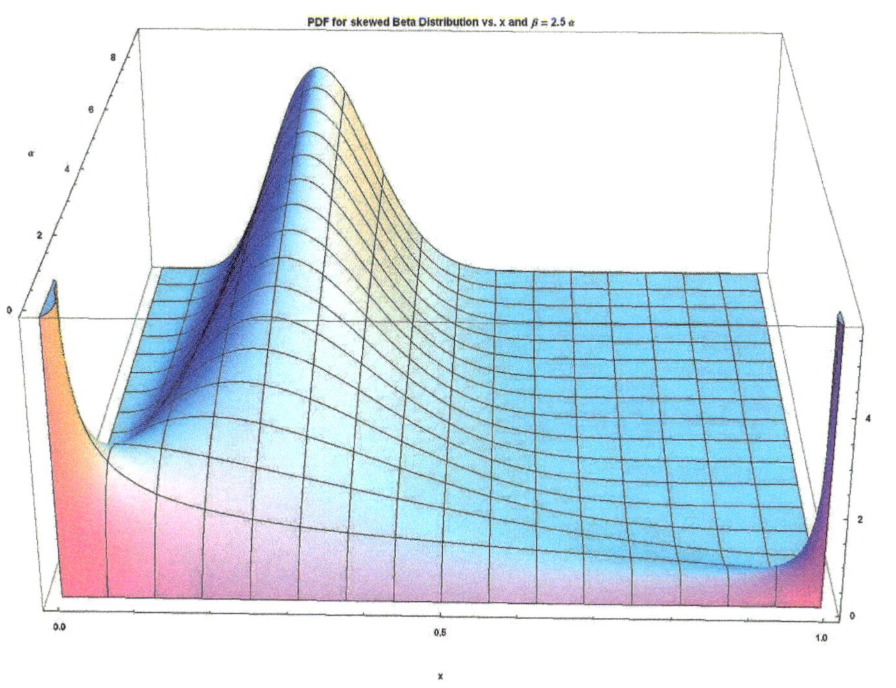

PDF for skewed beta distribution vs. x and beta= 2.5 alpha from 0 to 9

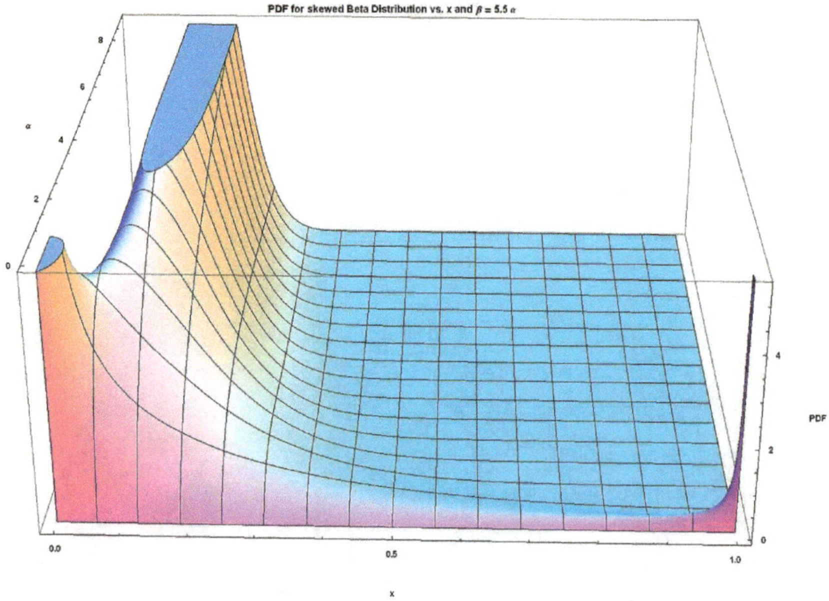

PDF for skewed beta distribution vs. x and beta= 5.5 alpha from 0 to 9

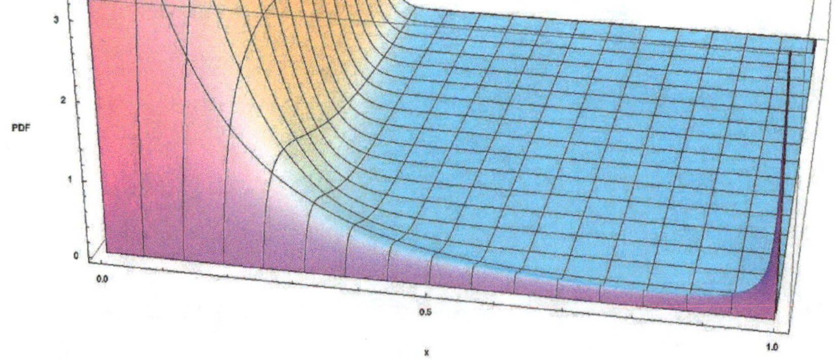

PDF for skewed beta distribution vs. x and beta= 8 alpha from 0 to 10.

The beta density function can take a wide variety of different shapes depending on the values of the two parameters α and β. The ability of the beta distribution to take this great diversity of shapes (using only two parameters) is partly responsible for finding wide application for modeling actual measurements:

Symmetric (α = β)

- the density function is symmetric about 1/2 (blue & teal plots).
- median = mean = 1/2.
- skewness = 0.
- **α = β < 1**
- U-shaped (blue plot).
- bimodal: left mode = 0, right mode =1, anti-mode = 1/2
- 1/12 < var(X) < 1/4
- −2 < excess kurtosis(X) < −6/5
- α = β = 1/2 is the arcsine distribution
- var(X) = 1/8
- excess kurtosis(X) = −3/2
- α = β → 0 is a 2-point Bernoulli distribution with equal probability 1/2 at each Dirac delta function end x = 0 and x = 1 and zero probability everywhere else. A coin toss: one face of the coin being x = 0 and the other face being x = 1.
- $\lim_{\alpha=\beta\to 0} \text{var}(X) = \frac{1}{4}$
- $\lim_{\alpha=\beta\to 0} \text{excess kurtosis}(X) = -2$ a lower value than this is impossible for any distribution to reach.
- The differential entropy approaches a minimum value of −∞
- α = β = 1
- the uniform [0, 1] distribution
- no mode
- var(X) = 1/12
- excess kurtosis(X) = −6/5
- The (negative anywhere else) differential entropy reaches its maximum value of zero
- α = β > 1
- symmetric unimodal
- mode = 1/2.
- 0 < var(X) < 1/12
- −6/5 < excess kurtosis(X) < 0
- α = β = 3/2 is a semi-elliptic [0, 1] distribution, see: Wigner semicircle distribution
- var(X) = 1/16.
- excess kurtosis(X) = −1
- α = β = 2 is the parabolic [0, 1] distribution
- var(X) = 1/20

- excess kurtosis(X) = −6/7
- $\alpha = \beta > 2$ is bell-shaped, with inflection points located to either side of the mode
- 0 < var(X) < 1/20
- −6/7 < excess kurtosis(X) < 0
- $\alpha = \beta \to \infty$ is a 1-point Degenerate distribution with a Dirac delta function spike at the midpoint x = 1/2 with probability 1, and zero probability everywhere else. There is 100% probability (absolute certainty) concentrated at the single point x = 1/2.
- $\lim_{\alpha=\beta\to\infty} \text{var}(X) = 0$
- $\lim_{\alpha=\beta\to\infty} \text{excess kurtosis}(X) = 0$
- The differential entropy approaches a minimum value of $-\infty$

Skewed (α ≠ β)

The density function is skewed. An interchange of parameter values yields the mirror image (the reverse) of the initial curve, some more specific cases:

- $\alpha < 1, \beta < 1$
  - U-shaped
  - Positive skew for $\alpha < \beta$, negative skew for $\alpha > \beta$.
  - bimodal: left mode = 0, right mode = 1, anti-mode = $\frac{\alpha-1}{\alpha+\beta-2}$
  - 0 < median < 1.
  - 0 < var(X) < 1/4
- $\alpha > 1, \beta > 1$
  - unimodal (magenta & cyan plots),
  - Positive skew for $\alpha < \beta$, negative skew for $\alpha > \beta$.
  - $\frac{\alpha-1}{\alpha+\beta-2}$
  - 0 < median < 1
  - 0 < var(X) < 1/12
- $\alpha < 1, \beta \geq 1$
  - reverse J-shaped with a right tail,
  - positively skewed,
  - strictly decreasing, convex
  - mode = 0
  - 0 < median < 1/2.
  - $0 < \text{var}(X) < \frac{-11+5\sqrt{5}}{2}$, (maximum variance occurs for $\alpha = 1, \beta = \frac{-1+\sqrt{5}}{2}$, or β = Φ the golden ratio conjugate)
- $\alpha \geq 1, \beta < 1$

- J-shaped with a left tail,
- negatively skewed,
- strictly increasing, convex
- mode = 1
- 1/2 < median < 1
- $0 < \mathrm{var}(X) < \frac{-11+5\sqrt{5}}{2}$, (maximum variance occurs for $\alpha = 1, \beta = \frac{-1+\sqrt{5}}{2}$, or β = Φ the golden ratio conjugate)

■ α = 1, β > 1

- positively skewed,
- strictly decreasing (red plot),
- a reversed (mirror-image) power function [0,1] distribution
- mode = 0
- α = 1, 1 < β < 2
    - concave
    - $1 - \frac{1}{\sqrt{2}} < \text{median} < \frac{1}{2}$
    - 1/18 < var(X) < 1/12.
- α = 1, β = 2
    - a straight line with slope −2, the right-triangular distribution with right angle at the left end, at x = 0
    - $\text{median} = 1 - \frac{1}{\sqrt{2}}$
    - var(X) = 1/18
- α = 1, β > 2
    - reverse J-shaped with a right tail,
    - convex
    - $0 < \text{median} < 1 - \frac{1}{\sqrt{2}}$
    - 0 < var(X) < 1/18

- **α > 1, β = 1**
    - negatively skewed,
    - strictly increasing (green plot),
    - the power function [0, 1] distribution

- mode =1
- $2 > \alpha > 1, \beta = 1$
    - concave
    - $0 < \text{median} < 1 - \frac{1}{\sqrt{2}}$
    - 1/18 < var(X) < 1/12
- $\alpha = 2, \beta = 1$
    - a straight line with slope +2, the right-triangular distribution with right angle at the right end, at x = 1
    - $\frac{1}{2} < \text{median} < \frac{1}{\sqrt{2}}$
    - var(X) = 1/18
- $\alpha > 2, \beta = 1$
    - J-shaped with a left tail, convex
    - $\frac{1}{\sqrt{2}} < \text{median} < 1$
    - 0 < var(X) < 1/18

Parameter Estimation

Method of Moments

Two Unknown Parameters

Two unknown parameters ($(\hat{\alpha}, \hat{\beta})$ of a beta distribution supported in the [0,1] interval) can be estimated, using the method of moments, with the first two moments (sample mean and sample variance) as follows. Let:

$$\text{sample mean}(X) = \bar{x} = \frac{1}{N}\sum_{i=1}^{N} X_i$$

be the sample mean estimate and

$$\text{sample variance}(X) = \bar{v} = \frac{1}{N-1}\sum_{i=1}^{N}(X_i - \bar{x})^2$$

be the sample variance estimate. The method-of-moments estimates of the parameters are

$$\hat{\alpha} = \bar{x}\left(\frac{\bar{x}(1-\bar{x})}{\bar{v}} - 1\right), \text{ if } \bar{v} < \bar{x}(1-\bar{x}),$$

$$\hat{\beta} = (1-\bar{x})\left(\frac{\bar{x}(1-\bar{x})}{\bar{v}} - 1\right), \text{ if } \bar{v} < \bar{x}(1-\bar{x}).$$

When the distribution is required over known interval other than [0, 1] with random variable X, say [a, c] with random variable Y, then replace $\bar{x}$ with $\frac{\bar{y}-a}{c-a}$, and $\bar{v}$ with $\frac{\bar{v}_Y}{(c-a)^2}$ in the above couple of equations for the shape parameters (see "Alternative parametrizations, four parameters" section below)., where:

$$\text{sample mean(Y)} = \bar{y} = \frac{1}{N}\sum_{i=1}^{N} Y_i$$

$$\text{sample variance(Y)} = \bar{v}_Y = \frac{1}{N-1}\sum_{i=1}^{N}(Y_i - \bar{y})^2$$

Four Unknown Parameters

Solutions for parameter estimates vs. (sample) excess Kurtosis and (sample) squared Skewness Beta distribution

All four parameters ($\hat{\alpha}, \hat{\beta}, \hat{a}, \hat{c}$ of a beta distribution supported in the [a, c] interval -see section "Alternative parametrizations, Four parameters"-) can be estimated, using the method of moments developed by Karl Pearson, by equating sample and population values of the first four central moments (mean, variance, skewness and excess kurtosis). The excess kurtosis was expressed in terms of the square of the skewness, and the sample size ν = α + β, (see previous section "Kurtosis") as follows:

$$\text{excess kurtosis} = \frac{6}{3+\nu}\left(\frac{(2+\nu)}{4}(\text{skewness})^2 - 1\right) \text{ if } (\text{skewness})^2 - 2 < \text{excess kurtosis} < \tfrac{3}{2}(\text{skewness})^2$$

One can use this equation to solve for the sample size ν= α + β in terms of the square of the skewness and the excess kurtosis as follows:

$$\hat{\nu} = \hat{\alpha} + \hat{\beta} = 3\frac{(\text{sample excess kurtosis}) - (\text{sample skewness})^2 + 2}{\frac{3}{2}(\text{sample skewness})^2 - (\text{sample excess kurtosis})}$$
$$\text{if } (\text{sample skewness})^2 - 2 < \text{sample excess kurtosis} < \tfrac{3}{2}(\text{sample skewness})^2$$

This is the ratio (multiplied by a factor of 3) between the previously derived limit boundaries for the beta distribution in a space (as originally done by Karl Pearson) defined with coordinates of the square of the skewness in one axis and the excess kurtosis in the other axis (see previous section titled "Kurtosis bounded by the square of the skewness"):

The case of zero skewness, can be immediately solved because for zero skewness, α = β and hence ν = 2α = 2β, therefore α = β = ν/2

$$\hat{\alpha} = \hat{\beta} = \frac{\hat{\nu}}{2} = \frac{\frac{3}{2}(\text{sample excess kurtosis}) + 3}{-(\text{sample excess kurtosis})}$$
$$\text{if sample skewness} = 0 \text{ and } -2 < \text{sample excess kurtosis} < 0$$

(Excess kurtosis is negative for the beta distribution with zero skewness, ranging from -2 to 0, so that $\hat{\nu}$ -and therefore the sample shape parameters- is positive, ranging from zero when the shape parameters approach zero and the excess kurtosis approaches -2, to infinity when the shape parameters approach infinity and the excess kurtosis approaches zero).

For non-zero sample skewness one needs to solve a system of two coupled equations. Since the skewness and the excess kurtosis are independent of the parameters $\hat{a}, \hat{c}$, the parameters $\hat{\alpha}, \hat{\beta}$ can be uniquely determined from the sample skewness and the sample excess kurtosis, by solving the coupled equations with two known variables (sample skewness and sample excess kurtosis) and two unknowns (the shape parameters):

$$(\text{sample skewness})^2 = \frac{4(\hat{\beta} - \hat{\alpha})^2(1 + \hat{\alpha} + \hat{\beta})}{\hat{\alpha}\hat{\beta}(2 + \hat{\alpha} + \hat{\beta})^2}$$

$$\text{sample excess kurtosis} = \frac{6}{3 + \hat{\alpha} + \hat{\beta}}\left(\frac{(2 + \hat{\alpha} + \hat{\beta})}{4}(\text{sample skewness})^2 - 1\right)$$

$$\text{if } (\text{sample skewness})^2 - 2 < \text{sample excess kurtosis} < \tfrac{3}{2}(\text{sample skewness})^2$$

resulting in the following solution:

$$\hat{\alpha}, \hat{\beta} = \frac{\hat{\nu}}{2}\left(1 \pm \frac{1}{\sqrt{1 + \frac{16(\hat{\nu}+1)}{(\hat{\nu}+2)^2(\text{sample skewness})^2}}}\right)$$

$$\text{if sample skewness} \neq 0 \text{ and } (\text{sample skewness})^2 - 2 < \text{sample excess kurtosis} < \tfrac{3}{2}(\text{sample skewness})^2$$

Where one should take the solutions as follows: $\hat{\alpha} > \hat{\beta}$ for (negative) sample skewness < 0, and $\hat{\alpha} < \hat{\beta}$ for (positive) sample skewness > 0.

The accompanying plot shows these two solutions as surfaces in a space with horizontal axes of (sample excess kurtosis) and (sample squared skewness) and the shape parameters as the vertical axis. The surfaces are constrained by the condition that the sample excess kurtosis must be bounded by the sample squared skewness as stipulated in the above equation. The two surfaces meet at the right edge defined by zero skewness. Along this right edge, both parameters are equal and the distribution is symmetric U-shaped for $\alpha = \beta < 1$, uniform for $\alpha = \beta = 1$, upside-down-U-shaped for $1 < \alpha = \beta < 2$ and bell-shaped for $\alpha = \beta > 2$. The surfaces also meet at the front (lower) edge defined by «the impossible boundary» line (excess kurtosis + 2 - skewness^2 = 0). Along this front (lower) boundary both shape parameters approach zero, and the probability density is concentrated more at one end than the other end (with practically nothing in between), with probabilities $p = \frac{\beta}{\alpha+\beta}$ at the left end x = 0 and $q = 1 - p = \frac{\alpha}{\alpha+\beta}$ at the right end x = 1. The two surfaces become further apart towards the rear edge. At this rear edge the surface parameters are quite different from each other. As remarked, for example, by Bowman and Shenton, sampling in the neighborhood of the line (sample excess kurtosis - (3/2)(sample skewness)^2 = 0) (the just-J-shaped portion of the rear edge where blue meets beige), "is dangerously near to chaos", because at that line the denominator of the expression above for the estimate $\nu = \alpha + \beta$ becomes zero and hence ν approaches infinity as that line is approached. Bowman and Shenton write that "the higher moment parameters (kurtosis and skewness) are extremely fragile (near that line). However the mean and standard deviation are fairly reliable." Therefore, the problem is for the case of four parameter estimation for very skewed distributions such that the excess kurtosis approaches (3/2) times the square of the skewness. This boundary line is produced by extremely skewed distributions with very large values of one of the parameters and very small values of the other parameter. See section titled "Kurtosis bounded by the square of the skewness" for a numerical example and further comments about this rear edge boundary line (sample excess kurtosis - (3/2) (sample skewness)^2 = 0). As remarked by Karl Pearson himself this issue may not be of much practical importance as this trouble arises only for very skewed J-shaped (or mirror-image J-shaped) distributions with very different values of shape parameters that are unlikely to occur much in practice). The usual skewed skewed-bell-shape distributions that occur in practice do not have this parameter estimation problem.

The remaining two parameters $\hat{a}, \hat{c}$ can be determined using the sample mean and the sample variance using a variety of equations. One alternative is to calculate the support interval range $(\hat{c} - \hat{a})$ based on the sample variance and the sample kurtosis. For this purpose one can solve, in terms of the range $(\hat{c} - \hat{a})$, the equation expressing the excess kurtosis in terms of the sample variance, and the sample size ν (see section titled "Kurtosis" and "Alternative parametrizations, four parameters"):

$$\text{sample excess kurtosis} = \frac{6}{(3+\hat{\nu})(2+\hat{\nu})}\left(\frac{(\hat{c}-\hat{a})^2}{(\text{sample variance})} - 6 - 5\hat{\nu}\right)$$

to obtain:

$$(\hat{c}-\hat{a}) = \sqrt{(\text{sample variance})}\sqrt{6 + 5\hat{\nu} + \frac{(2+\hat{\nu})(3+\hat{\nu})}{6}(\text{sample excess kurtosis})}$$

Another alternative is to calculate the support interval range $(\hat{c} - \hat{a})$ based on the sample variance and the sample skewness. For this purpose one can solve, in terms of the range $(\hat{c} - \hat{a})$, the equation expressing the squared skewness in terms of the sample variance, and the sample size ν (see section titled "Skewness" and "Alternative parametrizations, four parameters"):

$$(\text{sample skewness})^2 = \frac{4}{(2+\hat{\nu})^2}\left(\frac{(\hat{c}-\hat{a})^2}{(\text{sample variance})} - 4(1+\hat{\nu})\right)$$

to obtain:

$$(\hat{c}-\hat{a}) = \frac{\sqrt{(\text{sample variance})}}{2}\sqrt{(2+\hat{\nu})^2(\text{sample skewness})^2 + 16(1+\hat{\nu})}$$

The remaining parameter can be determined from the sample mean and the previously obtained parameters: $(\hat{c} - \hat{a}), \hat{\alpha}, \hat{\nu} = \hat{\alpha} + \hat{\beta}$:

$$\hat{a} = (\text{sample mean}) - \left(\frac{\hat{\alpha}}{\hat{\nu}}\right)(\hat{c}-\hat{a})$$

and finally, of course, $\hat{c} = (\hat{c} - \hat{a}) + \hat{a}$.

In the above formulas one may take, for example, as estimates of the sample moments:

$$\text{sample mean} = \bar{y} = \frac{1}{N}\sum_{i=1}^{N} Y_i$$

$$\text{sample variance} = \bar{v}_Y = \frac{1}{N-1}\sum_{i=1}^{N}(Y_i - \bar{y})^2$$

$$\text{sample skewness} = G_1 = \frac{N}{(N-1)(N-2)} \frac{\sum_{i=1}^{N}(Y_i - \bar{y})^3}{\bar{v}_Y^{\frac{3}{2}}}$$

$$\text{sample excess kurtosis} = G_2 = \frac{N(N+1)}{(N-1)(N-2)(N-3)} \frac{\sum_{i=1}^{N}(Y_i - \bar{y})^4}{\bar{v}_Y^2} - \frac{3(N-1)^2}{(N-2)(N-3)}$$

The estimators G_1 for sample skewness and G_2 for sample kurtosis are used by DAP/SAS, PSPP/SPSS, and Excel. However, they are not used by BMDP and (according to) they were not used by MINITAB in 1998. Actually, Joanes and Gill in their 1998 study concluded that the skewness and kurtosis estimators used in BMDP and in MINITAB (at that time) had smaller variance and mean-squared error in normal samples, but the skewness and kurtosis estimators used in DAP/SAS, PSPP/SPSS, namely G_1 and G_2, had smaller mean-squared error in samples from a very skewed distribution. It is for this reason that we have spelled out "sample skewness", etc., in the above formulas, to make it explicit that the user should choose the best estimator according to the problem at hand, as the best estimator for skewness and kurtosis depends on the amount of skewness (as shown by Joanes and Gill).

Maximum likelihood

Two Unknown Parameters

Max (Joint Log Likelihood/N) for Beta distribution Maxima at alpha=beta=2

Max (Joint Log Likelihood/N) for Beta distribution Maxima at alpha=beta={0.25,0.5,1,2,4,6,8}

As it is also the case for maximum likelihood estimates for the gamma distribution, the maximum likelihood estimates for the beta distribution do not have a general closed form solution for arbitrary values of the shape parameters. If $X_1, \ldots, X_N$ are independent random variables each having a beta distribution, the joint log likelihood function for N iid observations is:

$$\begin{aligned}
\ln \mathcal{L}(\alpha, \beta | X) &= \sum_{i=1}^{N} \ln(\mathcal{L}_i(\alpha, \beta | X_i)) \\
&= \sum_{i=1}^{N} \ln(f(X_i; \alpha, \beta)) \\
&= \sum_{i=1}^{N} \ln\left(\frac{X_i^{\alpha-1}(1-X_i)^{\beta-1}}{\mathrm{B}(\alpha, \beta)}\right) \\
&= (\alpha-1)\sum_{i=1}^{N} \ln(X_i) + (\beta-1)\sum_{i=1}^{N} \ln(1-X_i) - N \ln \mathrm{B}(\alpha, \beta)
\end{aligned}$$

Finding the maximum with respect to a shape parameter involves taking the partial derivative with respect to the shape parameter and setting the expression equal to zero yielding the maximum likelihood estimator of the shape parameters:

$$\frac{\partial \ln \mathcal{L}(\alpha,\beta|X)}{\partial \alpha} = \sum_{i=1}^{N} \ln X_i - N\frac{\partial \ln \mathrm{B}(\alpha,\beta)}{\partial \alpha} = 0$$
$$\frac{\partial \ln \mathcal{L}(\alpha,\beta|X)}{\partial \beta} = \sum_{i=1}^{N} \ln(1-X_i) - N\frac{\partial \ln \mathrm{B}(\alpha,\beta)}{\partial \beta} = 0$$

where:

$$\frac{\partial \ln \mathrm{B}(\alpha,\beta)}{\partial \alpha} = -\frac{\partial \ln \Gamma(\alpha+\beta)}{\partial \alpha} + \frac{\partial \ln \Gamma(\alpha)}{\partial \alpha} + \frac{\partial \ln \Gamma(\beta)}{\partial \alpha} = -\psi(\alpha+\beta) + \psi(\alpha) + 0$$
$$\frac{\partial \ln \mathrm{B}(\alpha,\beta)}{\partial \beta} = -\frac{\partial \ln \Gamma(\alpha+\beta)}{\partial \beta} + \frac{\partial \ln \Gamma(\alpha)}{\partial \beta} + \frac{\partial \ln \Gamma(\beta)}{\partial \beta} = -\psi(\alpha+\beta) + 0 + \psi(\beta)$$

since the **digamma function** denoted $\psi(\alpha)$ is defined as the logarithmic derivative of the gamma function:

$$\psi(\alpha) = \frac{\partial \ln \Gamma(\alpha)}{\partial \alpha}$$

To ensure that the values with zero tangent slope are indeed a maximum (instead of a saddle-point or a minimum) one has to also satisfy the condition that the curvature is negative. This amounts to satisfying that the second partial derivative with respect to the shape parameters is negative

$$\frac{\partial^2 \ln \mathcal{L}(\alpha,\beta|X)}{\partial \alpha^2} = -N\frac{\partial^2 \ln \mathrm{B}(\alpha,\beta)}{\partial \alpha^2} < 0$$
$$\frac{\partial^2 \ln \mathcal{L}(\alpha,\beta|X)}{\partial \beta^2} = -N\frac{\partial^2 \ln \mathrm{B}(\alpha,\beta)}{\partial \beta^2} < 0$$

using the previous equations, this is equivalent to:

$$\frac{\partial^2 \ln \mathrm{B}(\alpha,\beta)}{\partial \alpha^2} = \psi_1(\alpha) - \psi_1(\alpha+\beta) > 0$$
$$\frac{\partial^2 \ln \mathrm{B}(\alpha,\beta)}{\partial \beta^2} = \psi_1(\beta) - \psi_1(\alpha+\beta) > 0$$

where the **trigamma function**, denoted $\psi_1(\alpha)$, is the second of the polygamma functions, and is defined as the derivative of the digamma function:

$$\psi_1(\alpha) = \frac{\partial^2 \ln \Gamma(\alpha)}{\partial \alpha^2} = \frac{\partial \psi(\alpha)}{\partial \alpha}.$$

These conditions are equivalent to stating that the variances of the logarithmically transformed variables are positive, since:

$$\mathrm{var}[\ln(X)] = \mathrm{E}[\ln^2(X)] - (\mathrm{E}[\ln(X)])^2 = \psi_1(\alpha) - \psi_1(\alpha+\beta)$$
$$\mathrm{var}[\ln(1-X)] = \mathrm{E}[\ln^2(1-X)] - (\mathrm{E}[\ln(1-X)])^2 = \psi_1(\beta) - \psi_1(\alpha+\beta)$$

Therefore, the condition of negative curvature at a maximum is equivalent to the statements:

$$\mathrm{var}[\ln(X)] > 0$$
$$\mathrm{var}[\ln(1-X)] > 0$$

Alternatively, the condition of negative curvature at a maximum is also equivalent to stating that the following logarithmic derivatives of the geometric means G_X and $G_{(1-X)}$ are positive, since:

$$\psi_1(\alpha) - \psi_1(\alpha+\beta) = \frac{\partial \ln G_X}{\partial \alpha} > 0$$
$$\psi_1(\beta) - \psi_1(\alpha+\beta) = \frac{\partial \ln G_{(1-X)}}{\partial \beta} > 0$$

While these slopes are indeed positive, the other slopes are negative: $\frac{\partial \ln G_X}{\partial \beta}, \frac{\partial \ln G_{(1-X)}}{\partial \alpha} < 0.$

The slopes of the mean and the median with respect to α and β display similar sign behavior.

From the condition that at a maximum, the partial derivative with respect to the shape parameter equals zero, we obtain the following system of coupled maximum likelihood estimate equations (for the average log-likelihoods) that needs to be inverted to obtain the (unknown) shape parameter estimates $\hat{\alpha}, \hat{\beta}$ in terms of the (known) average of logarithms of the samples X_1, ..., X_N:

$$\hat{\mathrm{E}}[\ln(X)] = \psi(\hat{\alpha}) - \psi(\hat{\alpha}+\hat{\beta}) = \frac{1}{N}\sum_{i=1}^{N} \ln X_i = \ln \hat{G}_X$$
$$\hat{\mathrm{E}}[\ln(1-X)] = \psi(\hat{\beta}) - \psi(\hat{\alpha}+\hat{\beta}) = \frac{1}{N}\sum_{i=1}^{N} \ln(1-X_i) = \ln \hat{G}_{(1-X)}$$

where we recognize $\log \hat{G}_X$ as the logarithm of the sample geometric mean and $\log \hat{G}_{(1-X)}$ as the logarithm of the sample geometric mean based on (1-X), the mirror-image of X. For $\hat{\alpha} = \hat{\beta}$, it follows that $\hat{G}_X = \hat{G}_{(1-X)}$.

$$\hat{G}_X = \prod_{i=1}^{N} (X_i)^{\frac{1}{N}}$$
$$\hat{G}_{(1-X)} = \prod_{i=1}^{N} (1-X_i)^{\frac{1}{N}}$$

These coupled equations containing digamma functions of the shape parameter estimates $\hat{\alpha}, \hat{\beta}$ must be solved by numerical methods as done, for example, by Beckman et al. Gnanadesikan et al. give numerical solutions for a few cases.N.L.Johnson and S.Kotz suggest that for "not too small" shape parameter estimates $\hat{\alpha}, \hat{\beta}$, the logarithmic approximation to the digamma function $\psi(\hat{\alpha}) \approx \ln(\hat{\alpha} - \tfrac{1}{2})$ may be used to obtain initial values for an iterative solution, since the equations resulting from this approximation can be solved exactly:

$$\ln \frac{\hat{\alpha} - \frac{1}{2}}{\hat{\alpha}+\hat{\beta} - \frac{1}{2}} \approx \ln \hat{G}_X$$
$$\ln \frac{\hat{\beta} - \frac{1}{2}}{\hat{\alpha}+\hat{\beta} - \frac{1}{2}} \approx \ln \hat{G}_{(1-X)}$$

which leads to the following solution for the initial values (of the estimate shape parameters in terms of the sample geometric means) for an iterative solution:

$$\hat{\alpha} \approx \tfrac{1}{2} + \frac{\hat{G}_X}{2(1-\hat{G}_X-\hat{G}_{(1-X)})} \text{ if } \hat{\alpha} > 1$$

$$\hat{\beta} \approx \tfrac{1}{2} + \frac{\hat{G}_{(1-X)}}{2(1-\hat{G}_X-\hat{G}_{(1-X)})} \text{ if } \hat{\beta} > 1$$

Alternatively, the estimates provided by the method of moments can instead be used as initial values for an iterative solution of the maximum likelihood coupled equations in terms of the digamma functions.

When the distribution is required over a known interval other than [0, 1] with random variable X, say [a, c] with random variable Y, then replace $\ln(X_i)$ in the first equation with

$$\ln \frac{Y_i - a}{c - a},$$

and replace $\ln(1-X_i)$ in the second equation with

$$\ln \frac{c - Y_i}{c - a}$$

(see "Alternative parametrizations, four parameters" section below).

If one of the shape parameters is known, the problem is considerably simplified. The following logit transformation can be used to solve for the unknown shape parameter (for skewed cases such that $\hat{\alpha} \neq \hat{\beta}$, otherwise, if symmetric, both -equal- parameters are known when one is known):

$$\hat{\mathrm{E}}\left[\ln\left(\frac{X}{1-X}\right)\right] = \psi(\hat{\alpha}) - \psi(\hat{\beta}) = \frac{1}{N}\sum_{i=1}^{N} \ln \frac{X_i}{1-X_i} = \ln \hat{G}_X - \ln\left(\hat{G}_{(1-X)}\right)$$

This logit transformation is the logarithm of the transformation that divides the variable X by its mirror-image (X/(1 - X) resulting in the "inverted beta distribution" or beta prime distribution (also known as beta distribution of the second kind or Pearson›s Type VI) with support [0, +∞). As previously discussed in the section «Moments of logarithmically transformed random variables,» the logit transformation $\frac{X}{1-X}$, studied by Johnson, extends the finite support [0, 1] based on the original variable X to infinite support in both directions of the real line (−∞, +∞).

If, for example, $\hat{\beta}$ is known, the unknown parameter $\hat{\alpha}$ can be obtained in terms of the inverse digamma function of the right hand side of this equation:

$$\psi(\hat{\alpha}) = \frac{1}{N}\sum_{i=1}^{N} \ln \frac{X_i}{1-X_i} + \psi(\hat{\beta})$$

$$\hat{\alpha} = (\textbf{Inverse digamma})[\ln \hat{G}_X - \ln \hat{G}_{(1-X)} + \psi(\hat{\beta})]$$

In particular, if one of the shape parameters has a value of unity, for example for $\hat{\beta} = 1$ (the power function distribution with bounded support [0,1]), using the identity ψ(x + 1) = ψ(x) + 1/x in the equation ψ(x + 1) = ψ(x) + 1/x in the equation $\psi(\hat{\alpha}) - \psi(\hat{\alpha} + \hat{\beta}) = \ln \hat{G}_X$, the maximum likelihood estimator for the unknown parameter $\hat{\alpha}$ is, exactly:

$$\hat{\alpha} = -\frac{1}{\frac{1}{N}\sum_{i=1}^{N} \ln X_i} = -\frac{1}{\ln \hat{G}_X}$$

The beta has support [0, 1], therefore $\hat{G}_X < 1$, and hence $(-\ln \hat{G}_X) > 0$, and therefore $\hat{\alpha} > 0$.

In conclusion, the maximum likelihood estimates of the shape parameters of a beta distribution are (in general) a complicated function of the sample geometric mean, and of the sample geometric mean based on (1−X), the mirror-image of X. One may ask, if the variance (in addition to the mean) is necessary to estimate two shape parameters with the method of moments, why is the (logarithmic or geometric) variance not necessary to estimate two shape parameters with the maximum likelihood method, for which only the geometric means suffice? The answer is because the mean does not provide as much information as the geometric mean. For a beta distribution with equal shape parameters α = β, the mean is exactly 1/2, regardless of the value of the shape parameters, and therefore regardless of the value of the statistical dispersion (the variance). On the other hand, the geometric mean of a beta distribution with equal shape parameters α = β, depends on the value of the shape parameters, and therefore it contains more information. Also, the geometric mean of a beta distribution does not satisfy the symmetry conditions satisfied by the mean, therefore, by employing both the geometric mean based on X and geometric mean based on (1−X), the maximum likelihood method is able to provide best estimates for both parameters α = β, without need of employing the variance.

One can express the joint log likelihood per N iid observations in terms of the sufficient statistics (the sample geometric means) as follows:

$$\frac{\ln \mathcal{L}(\alpha, \beta|X)}{N} = (\alpha - 1)\ln \hat{G}_X + (\beta - 1)\ln \hat{G}_{(1-X)} - \ln \mathrm{B}(\alpha, \beta)$$

We can plot the joint log likelihood per N observations for fixed values of the sample geometric means to see the behavior of the likelihood function as a function of the shape parameters α and β. In such a plot, the shape parameter estimators $\hat{\alpha}, \hat{\beta}$ correspond to the maxima of the likelihood function. See the accompanying graph that shows that all the likelihood functions intersect at α = β = 1, which corresponds to the values of the shape parameters that give the maximum entropy (the maximum entropy occurs for shape parameters equal to unity: the uniform distribution). It is evident from the plot that the likelihood function gives sharp peaks for values of the shape parameter estimators close to zero, but that for values of the shape parameters estimators greater than one, the likelihood function becomes quite flat, with less defined peaks. Obviously, the maximum likelihood parameter estimation method for the beta distribution becomes less acceptable for larger values of the shape parameter estimators, as the uncertainty in the peak definition increases with the value of the shape parameter estimators. One can arrive at the same conclusion by noticing that the expression for the curvature of the likelihood function is in terms of the geometric variances

$$\frac{\partial^2 \ln \mathcal{L}(\alpha, \beta|X)}{\partial \alpha^2} = -\operatorname{var}[\ln X]$$

$$\frac{\partial^2 \ln \mathcal{L}(\alpha, \beta|X)}{\partial \beta^2} = -\operatorname{var}[\ln(1 - X)]$$

These variances (and therefore the curvatures) are much larger for small values of the shape parameter α and β. However, for shape parameter values α, β > 1, the variances (and therefore the curvatures) flatten out. Equivalently, this result follows from the Cramér–Rao bound, since the Fisher information matrix components for the beta distribution are these logarithmic variances. The Cramér–Rao bound states that the variance of any unbiased estimator $\hat{\alpha}$ of α is bounded by the reciprocal of the Fisher information:

$$\frac{\partial^2 \ln \mathcal{L}(\alpha,\beta|X)}{\partial\alpha^2} = -\operatorname{var}[\ln X]$$
$$\frac{\partial^2 \ln \mathcal{L}(\alpha,\beta|X)}{\partial\beta^2} = -\operatorname{var}[\ln(1-X)]$$

so the variance of the estimators increases with increasing α and β, as the logarithmic variances decrease.

Also one can express the joint log likelihood per N iid observations in terms of the digamma function expressions for the logarithms of the sample geometric means as follows:

$$\operatorname{var}(\hat{\alpha}) \geq \frac{1}{\operatorname{var}[\ln X]} \geq \frac{1}{\psi_1(\hat{\alpha}) - \psi_1(\hat{\alpha}+\hat{\beta})}$$
$$\operatorname{var}(\hat{\beta}) \geq \frac{1}{\operatorname{var}[\ln(1-X)]} \geq \frac{1}{\psi_1(\hat{\beta}) - \psi_1(\hat{\alpha}+\hat{\beta})}$$

this expression is identical to the negative of the cross-entropy (see section on "Quantities of information (entropy)"). Therefore, finding the maximum of the joint log likelihood of the shape parameters, per N iid observations, is identical to finding the minimum of the cross-entropy for the beta distribution, as a function of the shape parameters.

$$\frac{\ln \mathcal{L}(\alpha,\beta|X)}{N} = -H = -h - D_{\mathrm{KL}} = -\ln \mathrm{B}(\alpha,\beta) + (\alpha-1)\psi(\hat{\alpha}) + (\beta-1)\psi(\hat{\beta}) - (\alpha+\beta-2)\psi(\hat{\alpha}+\hat{\beta})$$

with the cross-entropy defined as follows:

$$H = \int_0^1 -f(X;\hat{\alpha},\hat{\beta}) \ln(f(X;\alpha,\beta))\,\mathrm{d}X$$

Four Unknown Parameters

The procedure is similar to the one followed in the two unknown parameter case. If $Y_1, \ldots, Y_N$ are independent random variables each having a beta distribution with four parameters, the joint log likelihood function for N iid observations is:

$$\begin{aligned}
\ln \mathcal{L}(\alpha,\beta,a,c|Y) &= \sum_{i=1}^{N} \ln \mathcal{L}_i(\alpha,\beta,a,c|Y_i) \\
&= \sum_{i=1}^{N} \ln f(Y_i;\alpha,\beta,a,c) \\
&= \sum_{i=1}^{N} \ln \frac{(Y_i-a)^{\alpha-1}(c-Y_i)^{\beta-1}}{(c-a)^{\alpha+\beta-1}\mathrm{B}(\alpha,\beta)} \\
&= (\alpha-1)\sum_{i=1}^{N}\ln(Y_i-a) + (\beta-1)\sum_{i=1}^{N}\ln(c-Y_i) - N\ln\mathrm{B}(\alpha,\beta) - N(\alpha+\beta-1)\ln(c-a)
\end{aligned}$$

Finding the maximum with respect to a shape parameter involves taking the partial derivative with respect to the shape parameter and setting the expression equal to zero yielding the maximum likelihood estimator of the shape parameters:

$$\frac{\partial \ln \mathcal{L}(\alpha,\beta,a,c|Y)}{\partial \alpha} = \sum_{i=1}^{N} \ln(Y_i - a) - N(-\psi(\alpha+\beta) + \psi(\alpha)) - N\ln(c-a) = 0$$
$$\frac{\partial \ln \mathcal{L}(\alpha,\beta,a,c|Y)}{\partial \beta} = \sum_{i=1}^{N} \ln(c - Y_i) - N(-\psi(\alpha+\beta) + \psi(\beta)) - N\ln(c-a) = 0$$
$$\frac{\partial \ln \mathcal{L}(\alpha,\beta,a,c|Y)}{\partial a} = -(\alpha-1)\sum_{i=1}^{N} \frac{1}{Y_i - a} + N(\alpha+\beta-1)\frac{1}{c-a} = 0$$
$$\frac{\partial \ln \mathcal{L}(\alpha,\beta,a,c|Y)}{\partial c} = (\beta-1)\sum_{i=1}^{N} \frac{1}{c - Y_i} - N(\alpha+\beta-1)\frac{1}{c-a} = 0$$

these equations can be re-arranged as the following system of four coupled equations (the first two equations are geometric means and the second two equations are the harmonic means) in terms of the maximum likelihood estimates for the four parameters $\hat{\alpha},\hat{\beta},\hat{a},\hat{c}$:

$$\frac{1}{N}\sum_{i=1}^{N} \ln\frac{Y_i - \hat{a}}{\hat{c} - \hat{a}} = \psi(\hat{\alpha}) - \psi(\hat{\alpha}+\hat{\beta}) = \ln \hat{G}_X$$
$$\frac{1}{N}\sum_{i=1}^{N} \ln\frac{\hat{c} - Y_i}{\hat{c} - \hat{a}} = \psi(\hat{\beta}) - \psi(\hat{\alpha}+\hat{\beta}) = \ln \hat{G}_{1-X}$$
$$\frac{1}{\frac{1}{N}\sum_{i=1}^{N}\frac{\hat{c}-\hat{a}}{Y_i-\hat{a}}} = \frac{\hat{\alpha}-1}{\hat{\alpha}+\hat{\beta}-1} = \hat{H}_X$$
$$\frac{1}{\frac{1}{N}\sum_{i=1}^{N}\frac{\hat{c}-\hat{a}}{\hat{c}-Y_i}} = \frac{\hat{\beta}-1}{\hat{\alpha}+\hat{\beta}-1} = \hat{H}_{1-X}$$

with sample geometric means:

$$\hat{G}_X = \prod_{i=1}^{N}\left(\frac{Y_i - \hat{a}}{\hat{c} - \hat{a}}\right)^{\frac{1}{N}}$$
$$\hat{G}_{(1-X)} = \prod_{i=1}^{N}\left(\frac{\hat{c} - Y_i}{\hat{c} - \hat{a}}\right)^{\frac{1}{N}}$$

The parameters $\hat{a},\hat{c}$ are embedded inside the geometric mean expressions in a nonlinear way (to the power 1/N). This precludes, in general, a closed form solution, even for an initial value approximation for iteration purposes. One alternative is to use as initial values for iteration the values obtained from the method of moments solution for the four parameter case. Furthermore, the expressions for the harmonic means are well-defined only for $\hat{\alpha},\hat{\beta} > 1$, which precludes a maximum likelihood solution for shape parameters less than unity in the four-parameter case. Fisher's information matrix for the four parameter case is positive-definite only for α, β > 2 (for further discussion, see section on Fisher information matrix, four parameter case), for bell-shaped (symmetric or unsymmetric) beta distributions, with inflection points located to either side of the mode. The following Fisher information components (that represent the expectations of the curvature of the log likelihood function) have singularities at the following values:

$$\alpha = 2: \quad \mathrm{E}\left[-\frac{1}{N}\frac{\partial^2 \ln \mathcal{L}(\alpha,\beta,a,c|Y)}{\partial a^2}\right] = \mathcal{I}_{a,a}$$

$$\beta = 2: \quad \mathrm{E}\left[-\frac{1}{N}\frac{\partial^2 \ln \mathcal{L}(\alpha,\beta,a,c|Y)}{\partial c^2}\right] = \mathcal{I}_{c,c}$$

$$\alpha = 2: \quad \mathrm{E}\left[-\frac{1}{N}\frac{\partial^2 \ln \mathcal{L}(\alpha,\beta,a,c|Y)}{\partial \alpha \partial a}\right] = \mathcal{I}_{\alpha,a}$$

$$\beta = 1: \quad \mathrm{E}\left[-\frac{1}{N}\frac{\partial^2 \ln \mathcal{L}(\alpha,\beta,a,c|Y)}{\partial \beta \partial c}\right] = \mathcal{I}_{\beta,c}$$

(for further discussion see section on Fisher information matrix). Thus, it is not possible to strictly carry on the maximum likelihood estimation for some well known distributions belonging to the four-parameter beta distribution family, like the uniform distribution (Beta(1, 1, a, c)), and the arcsine distribution (Beta(1/2, 1/2, a, c)). N.L.Johnson and S.Kotz ignore the equations for the harmonic means and instead suggest "If a and c are unknown, and maximum likelihood estimators of a, c, α and β are required, the above procedure (for the two unknown parameter case, with X transformed as X = (Y−a)/(c−a)) can be repeated using a succession of trial values of a and c, until the pair (a, c) for which maximum likelihood (given a and c) is as great as possible, is attained" (where, for the purpose of clarity, their notation for the parameters has been translated into the present notation).

Fisher Information Matrix

Let a random variable X have a probability density f(x;α). The partial derivative with respect to the (unknown, and to be estimated) parameter α of the log likelihood function is called the score. The second moment of the score is called the Fisher information:

$$\mathcal{I}(\alpha) = \mathrm{E}\left[\left(\frac{\partial}{\partial\alpha}\ln \mathcal{L}(\alpha|X)\right)^2\right],$$

The expectation of the score is zero, therefore the Fisher information is also the second moment centered on the mean of the score: the variance of the score.

If the log likelihood function is twice differentiable with respect to the parameter α, and under certain regularity conditions,then the Fisher information may also be written as follows (which is often a more convenient form for calculation purposes):

$$\mathrm{var}[\hat{\alpha}] \geq \frac{1}{\mathcal{I}(\alpha)}.$$

Thus, the Fisher information is the negative of the expectation of the second derivative with respect to the parameter α of the log likelihood function. Therefore, Fisher information is a measure of the curvature of the log likelihood function of α. A low curvature (and therefore high radius of curvature), flatter log likelihood function curve has low Fisher information; while a log likelihood function curve with large curvature (and therefore low radius of curvature) has high Fisher information. When the Fisher information matrix is computed at the evaluates of the parameters («the observed Fisher information matrix») it is equivalent to the replacement of the true log likelihood surface by a Taylor›s series approximation, taken as far as the quadratic terms. The word information, in the context of Fisher information, refers to information about the parameters. Information such as: estimation, sufficiency and properties of variances of estimators. The Cramér–Rao bound states that the inverse of the Fisher information is a lower bound on the variance of any estimator of a parameter α:

$$\operatorname{var}[\hat{\alpha}] \geq \frac{1}{\mathcal{I}(\alpha)}.$$

The precision to which one can estimate the estimator of a parameter α is limited by the Fisher Information of the log likelihood function. The Fisher information is a measure of the minimum error involved in estimating a parameter of a distribution and it can be viewed as a measure of the resolving power of an experiment needed to discriminate between two alternative hypothesis of a parameter.

When there are N parameters

$$\begin{bmatrix} \theta_1 \\ \theta_2 \\ \dots \\ \theta_N \end{bmatrix},$$

then the Fisher information takes the form of an N×N positive semidefinite symmetric matrix, the Fisher Information Matrix, with typical element:

$$(\mathcal{I}(\theta))_{i,j} = \mathrm{E}\left[\left(\frac{\partial}{\partial\theta_i}\ln\mathcal{L}\right)\left(\frac{\partial}{\partial\theta_j}\ln\mathcal{L}\right)\right]$$

Under certain regularity conditions, the Fisher Information Matrix may also be written in the following form, which is often more convenient for computation:

$$(\mathcal{I}(\theta))_{i,j} = -\mathrm{E}\left[\frac{\partial^2}{\partial\theta_i\,\partial\theta_j}\ln(\mathcal{L})\right].$$

With X_1, ..., X_N iid random variables, an N-dimensional "box" can be constructed with sides X_1, ..., X_N. Costa and Covershow that the (Shannon) differential entropy h(X) is related to the volume of the typical set (having the sample entropy close to the true entropy), while the Fisher information is related to the surface of this typical set.

Two Parameters

For X_1, ..., X_N independent random variables each having a beta distribution parametrized with shape parameters α and β, the joint log likelihood function for N iid observations is:

$$\ln(\mathcal{L}(\alpha,\beta|X)) = (\alpha-1)\sum_{i=1}^{N}\ln X_i + (\beta-1)\sum_{i=1}^{N}\ln(1-X_i) - N\ln\mathrm{B}(\alpha,\beta)$$

therefore the joint log likelihood function per N iid observations is:

$$\frac{1}{N}\ln(\mathcal{L}(\alpha,\beta|X)) = (\alpha-1)\frac{1}{N}\sum_{i=1}^{N}\ln X_i + (\beta-1)\frac{1}{N}\sum_{i=1}^{N}\ln(1-X_i) - \ln \mathrm{B}(\alpha,\beta)$$

For the two parameter case, the Fisher information has 4 components: 2 diagonal and 2 off-diagonal. Since the Fisher information matrix is symmetric, one of these off diagonal components is independent. Therefore, the Fisher information matrix has 3 independent components (2 diagonal and 1 off diagonal).

Aryal and Nadarajah calculated Fisher's information matrix for the four parameter case, from which the two parameter case can be obtained as follows:

$$-\frac{\partial^2 \ln \mathcal{L}(\alpha,\beta|X)}{N\partial\alpha^2} = \mathrm{var}[\ln(X)] = \psi_1(\alpha) - \psi_1(\alpha+\beta) = \mathcal{I}_{\alpha,\alpha} = \mathrm{E}\left[-\frac{\partial^2 \ln \mathcal{L}(\alpha,\beta|X)}{N\partial\alpha^2}\right] = \ln \mathrm{var}_{\mathrm{GX}}$$

$$-\frac{\partial^2 \ln \mathcal{L}(\alpha,\beta|X)}{N\partial\beta^2} = \mathrm{var}[\ln(1-X)] = \psi_1(\beta) - \psi_1(\alpha+\beta) = \mathcal{I}_{\beta,\beta} = \mathrm{E}\left[-\frac{\partial^2 \ln \mathcal{L}(\alpha,\beta|X)}{N\partial\beta^2}\right] = \ln \mathrm{var}_{\mathrm{G(1-X)}}$$

$$-\frac{\partial^2 \ln \mathcal{L}(\alpha,\beta|X)}{N\partial\alpha\partial\beta} = \mathrm{cov}[\ln X, \ln(1-X)] = -\psi_1(\alpha+\beta) = \mathcal{I}_{\alpha,\beta} = \mathrm{E}\left[-\frac{\partial^2 \ln \mathcal{L}(\alpha,\beta|X)}{N\partial\alpha\partial\beta}\right] = \ln \mathrm{cov}_{\mathrm{GX,(1-X)}}$$

Since the Fisher information matrix is symmetric

$$\mathcal{I}_{\alpha,\beta} = \mathcal{I}_{\beta,\alpha} = \ln \mathrm{cov}_{\mathrm{GX,(1-X)}}$$

The Fisher information components are equal to the log geometric variances and log geometric covariance. Therefore, they can be expressed as **trigamma functions**, denoted $\psi_1(\alpha)$, the second of the polygamma functions, defined as the derivative of the digamma function:

$$\psi_1(\alpha) = \frac{d^2 \ln \Gamma(\alpha)}{\partial\alpha^2} = \frac{\partial\psi(\alpha)}{\partial\alpha}.$$

These derivatives are also derived in the section titled "Parameter estimation", "Maximum likelihood", "Two unknown parameters," and plots of the log likelihood function are also shown in that section. The section titled "Geometric variance and covariance" contains plots and further discussion of the Fisher information matrix components: the log geometric variances and log geometric covariance as a function of the shape parameters α and β. The section titled "Other moments", "Moments of transformed random variables", "Moments of logarithmically transformed random variables" contains formulas for moments of logarithmically transformed random variables. Images for the Fisher information components $\mathcal{I}_{\alpha,\alpha}, \mathcal{I}_{\beta,\beta}$ and $\mathcal{I}_{\alpha,\beta}$ are shown in the section titled "Geometric variance".

The determinant of Fisher's information matrix is of interest (for example for the calculation of Jeffreys prior probability). From the expressions for the individual components of the Fisher information matrix, it follows that the determinant of Fisher's (symmetric) information matrix for the beta distribution is:

$$\begin{aligned}\det(\mathcal{I}(\alpha,\beta)) &= \mathcal{I}_{\alpha,\alpha}\mathcal{I}_{\beta,\beta} - \mathcal{I}_{\alpha,\beta}\mathcal{I}_{\alpha,\beta} \\ &= (\psi_1(\alpha) - \psi_1(\alpha+\beta))(\psi_1(\beta) - \psi_1(\alpha+\beta)) - (-\psi_1(\alpha+\beta))(-\psi_1(\alpha+\beta)) \\ &= \psi_1(\alpha)\psi_1(\beta) - (\psi_1(\alpha) + \psi_1(\beta))\psi_1(\alpha+\beta) \\ \lim_{\alpha\to 0}\det(\mathcal{I}(\alpha,\beta)) &= \lim_{\beta\to 0}\det(\mathcal{I}(\alpha,\beta)) = \infty \\ \lim_{\alpha\to\infty}\det(\mathcal{I}(\alpha,\beta)) &= \lim_{\beta\to\infty}\det(\mathcal{I}(\alpha,\beta)) = 0\end{aligned}$$

From Sylvester's criterion (checking whether the diagonal elements are all positive), it follows that the Fisher information matrix for the two parameter case is positive-definite (under the standard condition that the shape parameters are positive $\alpha>0$ and $\beta>0$).

Four Parameters

Fisher Information I(a,a) for $\alpha=\beta$ vs range (c-a) and exponent $\alpha=\beta$

Fisher Information I(α,a) for $\alpha=\beta$, vs. range (c - a) and exponent $\alpha=\beta$

If $Y_1, \ldots, Y_N$ are independent random variables each having a beta distribution with four parameters: the exponents α and β, as well as "a" (the minimum of the distribution range), and "c" (the maximum of the distribution range) (section titled "Alternative parametrizations", "Four parameters"), with probability density function:

$$f(y;\alpha,\beta,a,c)=\frac{f(x;\alpha,\beta)}{c-a}=\frac{\left(\frac{y-a}{c-a}\right)^{\alpha-1}\left(\frac{c-y}{c-a}\right)^{\beta-1}}{(c-a)B(\alpha,\beta)}=\frac{(y-a)^{\alpha-1}(c-y)^{\beta-1}}{(c-a)^{\alpha+\beta-1}B(\alpha,\beta)}.$$

the joint log likelihood function per N iid observations is:

$$\frac{1}{N}\ln(\mathcal{L}(\alpha,\beta,a,c|Y))=\frac{\alpha-1}{N}\sum_{i=1}^{N}\ln(Y_i-a)+\frac{\beta-1}{N}\sum_{i=1}^{N}\ln(c-Y_i)-\ln\mathrm{B}(\alpha,\beta)-(\alpha+\beta-1)\ln(c-a)$$

For the four parameter case, the Fisher information has 4*4=16 components. It has 12 off-diagonal components = (4*4 total - 4 diagonal). Since the Fisher information matrix is symmetric, half of these components (12/2=6) are independent. Therefore, the Fisher information matrix has 6 independent off-diagonal + 4 diagonal = 10 independent components. Aryal and Nadarajahcalculated Fisher's information matrix for the four parameter case as follows:

$$-\frac{1}{N}\frac{\partial^2\ln\mathcal{L}(\alpha,\beta,a,c|Y)}{\partial\alpha^2}=\mathrm{var}[\ln(X)]=\psi_1(\alpha)-\psi_1(\alpha+\beta)=\mathcal{I}_{\alpha,\alpha}=\mathrm{E}\left[-\frac{1}{N}\frac{\partial^2\ln\mathcal{L}(\alpha,\beta,a,c|Y)}{\partial\alpha^2}\right]=\ln(\mathrm{var}_{GX})$$

$$-\frac{1}{N}\frac{\partial^2\ln\mathcal{L}(\alpha,\beta,a,c|Y)}{\partial\beta^2}=\mathrm{var}[\ln(1-X)]=\psi_1(\beta)-\psi_1(\alpha+\beta)=\mathcal{I}_{\beta,\beta}=\mathrm{E}\left[-\frac{1}{N}\frac{\partial^2\ln\mathcal{L}(\alpha,\beta,a,c|Y)}{\partial\beta^2}\right]=\ln(\mathrm{var}_{G(1-X)})$$

$$-\frac{1}{N}\frac{\partial^2\ln\mathcal{L}(\alpha,\beta,a,c|Y)}{\partial\alpha\partial\beta}=\mathrm{cov}[\ln X,(1-X)]=-\psi_1(\alpha+\beta)=\mathcal{I}_{\alpha,\beta}=\mathrm{E}\left[-\frac{1}{N}\frac{\partial^2\ln\mathcal{L}(\alpha,\beta,a,c|Y)}{\partial\alpha\partial\beta}\right]=\ln(\mathrm{cov}_{GX,(1-X)})$$

In the above expressions, the use of X instead of Y in the expressions var[ln(X)] = ln(var_{GX}) is not an error. The expressions in terms of the log geometric variances and log geometric covariance occur as functions of the two parameter X ~ Beta(α, β) parametrization because when taking the partial derivatives with respect to the exponents (α, β) in the four parameter case, one obtains the identical expressions as for the two parameter case: these terms of the four parameter Fisher information matrix are independent of the minimum "a" and maximum "c" of the distribution's range. The only non-zero term upon double differentiation of the log likelihood function with respect to the exponents α and β is the second derivative of the log of the beta function: ln(B(α, β)). This term is independent of the minimum "a" and maximum "c" of the distribution's range. Double differentiation of this term results in trigamma functions. The sections titled "Maximum likelihood", "Two unknown parameters" and "Four unknown parameters" also show this fact.

The Fisher information for N i.i.d. samples is N times the individual Fisher information (eq. 11.279, page 394 of Cover and Thomas). (Aryal and Nadarajah take a single observation, N = 1, to calculate the following components of the Fisher information, which leads to the same result as considering the derivatives of the log likelihood per N observations. Moreover, below the erroneous expression for $\mathcal{I}_{a,a}$ in Aryal and Nadarajah has been corrected.)

$$\alpha > 2: \quad \mathrm{E}\left[-\frac{1}{N}\frac{\partial^2 \ln \mathcal{L}(\alpha,\beta,a,c|Y)}{\partial a^2}\right] = \mathcal{I}_{a,a} = \frac{\beta(\alpha+\beta-1)}{(\alpha-2)(c-a)^2}$$

$$\beta > 2: \quad \mathrm{E}\left[-\frac{1}{N}\frac{\partial^2 \ln \mathcal{L}(\alpha,\beta,a,c|Y)}{\partial c^2}\right] = \mathcal{I}_{c,c} = \frac{\alpha(\alpha+\beta-1)}{(\beta-2)(c-a)^2}$$

$$\mathrm{E}\left[-\frac{1}{N}\frac{\partial^2 \ln \mathcal{L}(\alpha,\beta,a,c|Y)}{\partial a \partial c}\right] = \mathcal{I}_{a,c} = \frac{(\alpha+\beta-1)}{(c-a)^2}$$

$$\alpha > 1: \quad \mathrm{E}\left[-\frac{1}{N}\frac{\partial^2 \ln \mathcal{L}(\alpha,\beta,a,c|Y)}{\partial \alpha \partial a}\right] = \mathcal{I}_{\alpha,a} = \frac{\beta}{(\alpha-1)(c-a)}$$

$$\mathrm{E}\left[-\frac{1}{N}\frac{\partial^2 \ln \mathcal{L}(\alpha,\beta,a,c|Y)}{\partial \alpha \partial c}\right] = \mathcal{I}_{\alpha,c} = \frac{1}{(c-a)}$$

$$\mathrm{E}\left[-\frac{1}{N}\frac{\partial^2 \ln \mathcal{L}(\alpha,\beta,a,c|Y)}{\partial \beta \partial a}\right] = \mathcal{I}_{\beta,a} = -\frac{1}{(c-a)}$$

$$\beta > 1: \quad \mathrm{E}\left[-\frac{1}{N}\frac{\partial^2 \ln \mathcal{L}(\alpha,\beta,a,c|Y)}{\partial \beta \partial c}\right] = \mathcal{I}_{\beta,c} = -\frac{\alpha}{(\beta-1)(c-a)}$$

The lower two diagonal entries of the Fisher information matrix, with respect to the parameter "a" (the minimum of the distribution's range): $\mathcal{I}_{a,a}$, and with respect to the parameter "c" (the maximum of the distribution's range): $\mathcal{I}_{c,c}$ are only defined for exponents α > 2 and β > 2 respectively. The Fisher information matrix component $\mathcal{I}_{a,a}$ for the minimum "a" approaches infinity for exponent α approaching 2 from above, and the Fisher information matrix component $\mathcal{I}_{c,c}$ for the maximum "c" approaches infinity for exponent β approaching 2 from above.

The Fisher information matrix for the four parameter case does not depend on the individual values of the minimum «a» and the maximum «c», but only on the total range (c−a). Moreover, the components of the Fisher information matrix that depend on the range (c−a), depend only through its inverse (or the square of the inverse), such that the Fisher information decreases for increasing range (c−a).

The accompanying images show the Fisher information components $\mathcal{I}_{a,a}$ and $\mathcal{I}_{\alpha,a}$. Images for the Fisher information components $\mathcal{I}_{\alpha,\alpha}$ and $\mathcal{I}_{\beta,\beta}$ are shown in the section titled "Geometric variance". All these Fisher information components look like a basin, with the "walls" of the basin being located at low values of the parameters.

The following four-parameter-beta-distribution Fisher information components can be expressed in terms of the two-parameter: X ~ Beta(α, β) expectations of the transformed ratio ((1-X)/X) and of its mirror image (X/(1-X)), scaled by the range (c−a), which may be helpful for interpretation:

$$\mathcal{I}_{\alpha,a} = \frac{\mathrm{E}\left[\frac{1-X}{X}\right]}{c-a} = \frac{\beta}{(\alpha-1)(c-a)} \text{ if } \alpha > 1$$

$$\mathcal{I}_{\beta,c} = -\frac{\mathrm{E}\left[\frac{X}{1-X}\right]}{c-a} = -\frac{\alpha}{(\beta-1)(c-a)} \text{ if } \beta > 1$$

These are also the expected values of the "inverted beta distribution" or beta prime distribution (also known as beta distribution of the second kind or Pearson›s Type VI) and its mirror image, scaled by the range (c−a).

Also, the following Fisher information components can be expressed in terms of the harmonic (1/X) variances or of variances based on the ratio transformed variables ((1-X)/X) as follows:

$$\alpha > 2: \quad \mathcal{I}_{a,a} = \operatorname{var}\left[\frac{1}{X}\right]\left(\frac{\alpha-1}{c-a}\right)^2 = \operatorname{var}\left[\frac{1-X}{X}\right]\left(\frac{\alpha-1}{c-a}\right)^2 = \frac{\beta(\alpha+\beta-1)}{(\alpha-2)(c-a)^2}$$

$$\beta > 2: \quad \mathcal{I}_{c,c} = \operatorname{var}\left[\frac{1}{1-X}\right]\left(\frac{\beta-1}{c-a}\right)^2 = \operatorname{var}\left[\frac{X}{1-X}\right]\left(\frac{\beta-1}{c-a}\right)^2 = \frac{\alpha(\alpha+\beta-1)}{(\beta-2)(c-a)^2}$$

$$\mathcal{I}_{a,c} = \operatorname{cov}\left[\frac{1}{X}, \frac{1}{1-X}\right]\frac{(\alpha-1)(\beta-1)}{(c-a)^2} = \operatorname{cov}\left[\frac{1-X}{X}, \frac{X}{1-X}\right]\frac{(\alpha-1)(\beta-1)}{(c-a)^2} = \frac{(\alpha+\beta-1)}{(c-a)^2}$$

See section "Moments of linearly transformed, product and inverted random variables" for these expectations.

The determinant of Fisher's information matrix is of interest (for example for the calculation of Jeffreys prior probability). From the expressions for the individual components, it follows that the determinant of Fisher's (symmetric) information matrix for the beta distribution with four parameters is:

$$\begin{aligned}
\det(\mathcal{I}(\alpha,\beta,a,c)) = & -\mathcal{I}_{a,c}^2\mathcal{I}_{\alpha,a}\mathcal{I}_{\alpha,\beta} + \mathcal{I}_{a,a}\mathcal{I}_{a,c}\mathcal{I}_{\alpha,c}\mathcal{I}_{\alpha,\beta} + \mathcal{I}_{a,c}^2\mathcal{I}_{\alpha,\beta}^2 - \mathcal{I}_{a,a}\mathcal{I}_{c,c}\mathcal{I}_{\alpha,\beta}^2 \\
& - \mathcal{I}_{a,c}\mathcal{I}_{\alpha,a}\mathcal{I}_{\alpha,c}\mathcal{I}_{\beta,a} + \mathcal{I}_{a,c}^2\mathcal{I}_{\alpha,\alpha}\mathcal{I}_{\beta,a} + 2\mathcal{I}_{c,c}\mathcal{I}_{\alpha,a}\mathcal{I}_{\alpha,\beta}\mathcal{I}_{\beta,a} \\
& - 2\mathcal{I}_{a,c}\mathcal{I}_{\alpha,c}\mathcal{I}_{\alpha,\beta}\mathcal{I}_{\beta,a} + \mathcal{I}_{\alpha,c}^2\mathcal{I}_{\beta,a}^2 - \mathcal{I}_{c,c}\mathcal{I}_{\alpha,\alpha}\mathcal{I}_{\beta,a}^2 + \mathcal{I}_{a,c}\mathcal{I}_{\alpha,a}^2\mathcal{I}_{\beta,c} \\
& - \mathcal{I}_{a,a}\mathcal{I}_{a,c}\mathcal{I}_{\alpha,\alpha}\mathcal{I}_{\beta,c} - \mathcal{I}_{a,c}\mathcal{I}_{\alpha,a}\mathcal{I}_{\alpha,\beta}\mathcal{I}_{\beta,c} + \mathcal{I}_{a,a}\mathcal{I}_{\alpha,c}\mathcal{I}_{\alpha,\beta}\mathcal{I}_{\beta,c} \\
& - \mathcal{I}_{\alpha,a}\mathcal{I}_{\alpha,c}\mathcal{I}_{\beta,a}\mathcal{I}_{\beta,c} + \mathcal{I}_{a,c}\mathcal{I}_{\alpha,\alpha}\mathcal{I}_{\beta,a}\mathcal{I}_{\beta,c} - \mathcal{I}_{c,c}\mathcal{I}_{\alpha,a}^2\mathcal{I}_{\beta,\beta} \\
& + 2\mathcal{I}_{a,c}\mathcal{I}_{\alpha,a}\mathcal{I}_{\alpha,c}\mathcal{I}_{\beta,\beta} - \mathcal{I}_{a,a}\mathcal{I}_{\alpha,c}^2\mathcal{I}_{\beta,\beta} - \mathcal{I}_{a,c}^2\mathcal{I}_{\alpha,\alpha}\mathcal{I}_{\beta,\beta} + \mathcal{I}_{a,a}\mathcal{I}_{c,c}\mathcal{I}_{\alpha,\alpha}\mathcal{I}_{\beta,\beta} \text{ if } \alpha,\beta > 2
\end{aligned}$$

Using Sylvester's criterion (checking whether the diagonal elements are all positive), and since diagonal components $\mathcal{I}_{a,a}$ and $\mathcal{I}_{c,c}$ have singularities at α=2 and β=2 it follows that the Fisher information matrix for the four parameter case is positive-definite for α>2 and β>2. Since for α > 2 and β > 2 the beta distribution is (symmetric or unsymmetric) bell shaped, it follows that the Fisher information matrix is positive-definite only for bell-shaped (symmetric or unsymmetric) beta distributions, with inflection points located to either side of the mode. Thus, important well known distributions belonging to the four-parameter beta distribution family, like the parabolic distribution (Beta(2,2,a,c)) and the uniform distribution (Beta(1,1,a,c)) have Fisher information components $(\mathcal{I}_{a,a}, \mathcal{I}_{c,c}, \mathcal{I}_{\alpha,a}, \mathcal{I}_{\beta,c})$ that blow up (approach infinity) in the four-parameter case (although their Fisher information components are all defined for the two parameter case). The four-parameter Wigner semicircle distribution (Beta(3/2,3/2,a,c)) and arcsine distribution (Beta(1/2,1/2,a,c)) have negative Fisher information determinants for the four-parameter case.

Generating beta-distributed random variates

If X and Y are independent, with $X \sim \Gamma(\alpha, \theta)$ and $Y \sim \Gamma(\beta, \theta)$ then

$\frac{X}{X+Y} \sim \mathrm{B}(\alpha, \beta)$.

So one algorithm for generating beta variates is to generate X/(X + Y), where X is a gamma variate with parameters (α, 1) and Y is an independent gamma variate with parameters (β, 1).

Also, the kth order statistic of n uniformly distributed variates is $\mathrm{B}(k, n+1-k)$, so an alternative if α and β are small integers is to generate α + β − 1 uniform variates and choose the α-th smallest.

Another way to generate the Beta distribution is by Pólya urn model. According to this method, one start with an "urn" with α "black" balls and β "white" balls and draw uniformly with replacement. Every trial an additional ball is added according to the color of the last ball which was drawn. Asymptotically, the proportion of black and white balls will be distributed according to the Beta distribution, where each repetition of the experiment will produce a different value.

Related Distributions

Transformations

- If X ~ Beta(α, β) then 1−X ~ Beta(β, α) mirror-image symmetry
- If X ~ Beta(α, β) then $\frac{X}{1-X} \sim \beta'(\alpha, \beta)$. The beta prime distribution, also called "beta distribution of the second kind".
- If X ~ Beta(n/2, m/2) then $\frac{mX}{n(1-X)} \sim F(n, m)$ (assuming n > 0 and m > 0), the Fisher-Snedecor F distribution.
- If $X \sim Beta\left(1 + \lambda\frac{m-\min}{\max - \min}, 1 + \lambda\frac{\max - m}{\max - \min}\right)$ then min+X(max−min) ~ PERT(min, max, m, λ) where PERT denotes a distribution used in PERT analysis, and m=most likely value. Traditionally λ = 4 in PERT analysis.
- If X ~ Beta(1, β) then X ~ Kumaraswamy distribution with parameters (1, β)
- If X ~ Beta(α, 1) then X ~ Kumaraswamy distribution with parameters (α, 1)
- If X ~ Beta(α, 1) then −ln(X) ~ Exponential(α)

Special and Limiting Cases

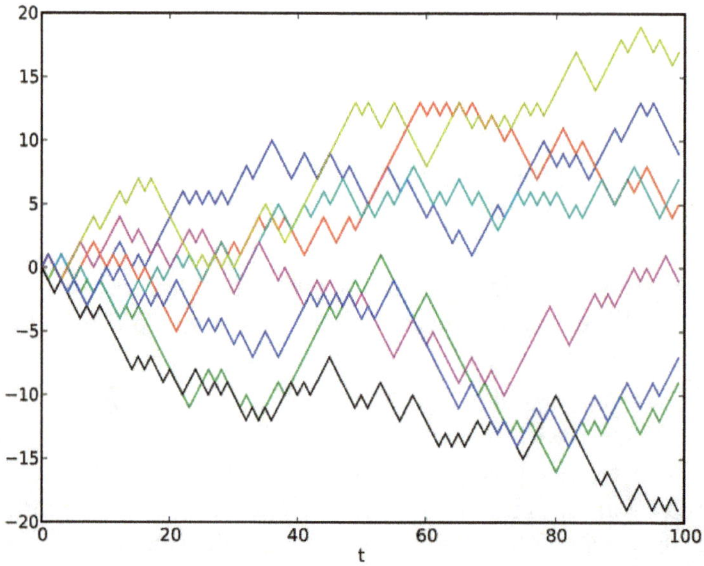

Example of eight realizations of a random walk in one dimension starting at 0: the probability for the time of the last visit to the origin is distributed as Beta(1/2, 1/2).

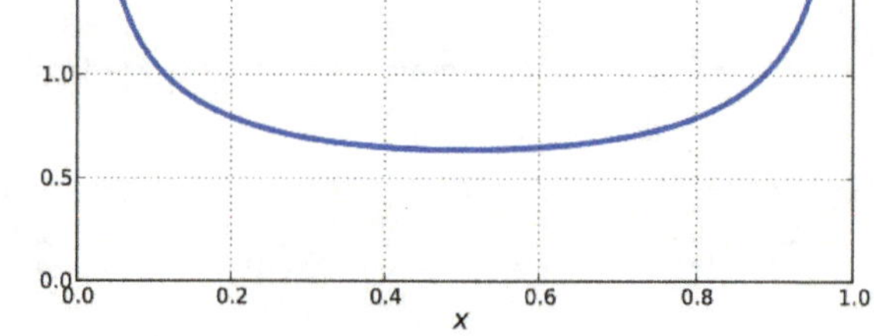

Beta(1/2, 1/2): The arcsine distribution probability density was proposed by Harold Jeffreys to represent uncertainty for a Bernoulli or a binomial distribution in Bayesian inference, and is now commonly referred to as Jeffreys prior: $p^{-1/2}(1-p)^{-1/2}$. This distribution also appears in several random walk fundamental theorems.

- Beta(1, 1) ~ U(0, 1).
- If X ~ Beta(3/2, 3/2) and r > 0 then 2rX−r ~ Wigner semicircle distribution.
- Beta(1/2, 1/2) is equivalent to the arcsine distribution. This distribution is also Jeffreys prior probability for the Bernoulli and binomial distributions . The arcsine probability density is a distribution that appears in several random walk fundamental theorems. In a fair coin toss random walk, the probability for the time of the last visit to the origin is distributed as an (U-shaped) arcsine distribution. In a two-player fair-coin-toss game, a player is said to be in the lead if the random walk (that started at the origin) is above the origin. The most probable number of times that a given player will be in the lead, in a game of length 2N, is not N. On the contrary, N is the least likely number of times that the player will be in the lead. The most likely number of times in the lead is 0 or 2N (following the arcsine distribution).
- $\lim_{n\to\infty} nBeta(1,n) = \text{Exponential}(1)$ the exponential distribution
- $\lim_{n\to\infty} nBeta(k,n) = \text{Gamma}(k,1)$ the gamma distribution

Derived from other Distributions

- The kth order statistic of a sample of size n from the uniform distribution is a beta random variable, $U_{(k)}$ ~ Beta(k, n+1−k).
- If X ~ Gamma(α, θ) and Y ~ Gamma(β, θ) are independent, then $\frac{X}{X+Y} \sim Beta(\alpha,\beta)$

  .
- If $X \sim \chi^2(\alpha)$ and $Y \sim \chi^2(\beta)$ are independent, then $\frac{X}{X+Y} \sim Beta(\frac{\alpha}{2},\frac{\beta}{2})$.
- If X ~ U(0, 1) and α > 0 then $X^{1/\alpha}$ ~ Beta(α, 1). The power function distribution.

Combination with other Distributions

- X ~ Beta(α, β) and Y ~ F(2β,2α) then {\displaystyle \Pr(X\leq {\tfrac {\alpha }{\alpha +\beta x}})=\Pr(Y\geq x)\,} for all x > 0.

Compounding with other distributions

- If p ~ Beta(α, β) and X ~ Bin(k, p) then X ~ beta-binomial distribution
- If p ~ Beta(α, β) and X ~ NB(r, p) then X ~ beta negative binomial distribution

Generalisations

- The Dirichlet distribution is a multivariate generalization of the beta distribution. Univariate marginals of the Dirichlet distribution have a beta distribution. The beta distribution is conjugate to the binomial and Bernoulli distributions in exactly the same way as the Dirichlet distribution is conjugate to the multinomial distribution and categorical distribution.

- The Pearson type I distribution is identical to the beta distribution (except for arbitrary shifting and re-scaling that can also be accomplished with the four parameter parametrization of the beta distribution).
- $Beta(\alpha,\beta) = \lim_{\delta\to 0} \text{NonCentralBeta}(\alpha,\beta,\delta)$ the noncentral beta distribution
- the Generalized beta distribution is a five-parameter distribution family which has the beta distribution as a special case.

REFERENCES

1. Pierre Simon de Laplace (1812). Analytical Theory of Probability.
2. The first major treatise blending calculus with probability theory, originally in French: Théorie Analytique des Probabilités.
3. A. Kolmogoroff (1933). Grundbegriffe der Wahrscheinlichkeitsrechnung. doi:10.1007/978-3-642-49888-6. ISBN 978-3-642-49888-6.
4. An English translation by Nathan Morrison appeared under the title Foundations of the Theory of Probability (Chelsea, New York) in 1950, with a second edition in 1956.
5. Patrick Billingsley (1979). Probability and Measure. New York, Toronto, London: John Wiley and Sons.
6. Olav Kallenberg; Foundations of Modern Probability, 2nd ed. Springer Series in Statistics. (2002). 650 pp. ISBN 0-387-95313-2
7. Henk Tijms (2004). Understanding Probability. Cambridge Univ. Press.
8. A lively introduction to probability theory for the beginner.
9. Olav Kallenberg; Probabilistic Symmetries and Invariance Principles. Springer -Verlag, New York (2005). 510 pp. ISBN 0-387-25115-4
10. Gut, Allan (2005). Probability: A Graduate Course. Springer-Verlag. ISBN 0-387-22833-0.

CHAPTER 6

Abstract Algebra: Concepts and Applications

Chapter Contents

- **Introduction**
- **Basic Concepts**
- **Algebraic Structure**
- **Applications**

Abstract algebra is the set of advanced topics of algebra that deal with abstract algebraic structures rather than the usual number systems. The most important of these structures are groups, rings, and fields. Important branches of abstract algebra are commutative algebra, representation theory, and homological algebra.

INTRODUCTION

In algebra, which is a broad division of mathematics, **abstract algebra** (occasionally called **modern algebra**) is the study of algebraic structures. Algebraic structures include groups, rings, fields, modules, vector spaces, lattices, and algebras. The term *abstract algebra* was coined in the early 20th century to distinguish this area of study from the other parts of algebra.

The permutations of Rubik›s Cube form a group, a fundamental concept within abstract algebra.

Algebraic structures, with their associated homomorphisms, form mathematical categories. Category theory is a formalism that allows a unified way for expressing properties and constructions that are similar for various structures.

Universal algebra is a related subject that studies types of algebraic structures as single objects. For example, the structure of groups is a single object in universal algebra, which is called *variety of groups*.

As in other parts of mathematics, concrete problems and examples have played important roles in the development of abstract algebra. Through the end of the nineteenth century, many -- perhaps most -- of these problems were in some way related to the theory of algebraic equations. Major themes include:

- Solving of systems of linear equations, which led to linear algebra
- Attempts to find formulae for solutions of general polynomial equations of higher degree that resulted in discovery of groups as abstract manifestations of symmetry
- Arithmetical investigations of quadratic and higher degree forms and diophantine equations, that directly produced the notions of a ring and ideal.

Numerous textbooks in abstract algebra start with axiomatic definitions of various algebraic structures and then proceed to establish their properties. This creates a false impression that in algebra axioms had come first and then served as a motivation and as a basis of further study. The true order of historical development was almost exactly the opposite. For example, the hypercomplex numbers of the nineteenth century had kinematic and physical motivations but challenged comprehension. Most theories that are now recognized as parts of algebra started as collections of disparate facts from various branches of mathematics, acquired a common theme that served as a core around which various results were grouped, and finally became unified on a basis of a common set of concepts. An archetypical example of this progressive synthesis can be seen in the history of group theory.

Early Group Theory

There were several threads in the early development of group theory, in modern language loosely corresponding to *number theory*, *theory of equations*, and *geometry*.

Leonhard Euler considered algebraic operations on numbers modulo an integer, modular arithmetic, in his generalization of Fermat›s little theorem. These investigations were taken much further by Carl Friedrich Gauss, who considered the structure of multiplicative groups of residues mod n and established many properties of cyclic and more general abelian groups that arise in this way. In his investigations of composition of binary quadratic forms, Gauss explicitly stated the associative law for the composition of forms, but like Euler before him, he seems to have been more interested in concrete results than in general theory. In 1870, Leopold Kronecker gave a definition of an abelian group in the context of ideal class groups of a number field, generalizing Gauss's work; but it appears he did not tie his definition with previous work on groups, particularly permutation groups. In 1882, considering the same question, Heinrich M. Weber realized the connection and gave a similar definition that involved the cancellation property but omitted the existence of the inverse element, which was sufficient in his context (finite groups).

Permutations were studied by Joseph-Louis Lagrange in his 1770 paper *Réflexions sur la résolution algébrique des équations (Thoughts on the algebraic solution of equations)* devoted to solutions of algebraic equations, in which he introduced Lagrange resolvents. Lagrange›s goal was to understand why equations of third and fourth degree admit formulae for solutions, and he identified as key objects permutations of the roots. An important novel step taken by Lagrange in this paper was the abstract view of the roots, i.e. as symbols and not as numbers. However, he did not consider composition of permutations. Serendipitously, the first edition of Edward Waring›s *Meditationes Algebraicae* (*Meditations on Algebra*) appeared in the same year, with an expanded version published in 1782. Waring proved the main theorem on symmetric functions, and specially considered the relation between the roots of a quartic equation and its resolvent cubic. *Mémoire sur la résolution des équations* (*Memoire on the Solving of Equations*) of Alexandre Vandermonde (1771) developed the theory of symmetric functions from a slightly different angle, but like Lagrange, with the goal of understanding solvability of algebraic equations.

Kronecker claimed in 1888 that the study of modern algebra began with this first paper of Vandermonde. Cauchy states quite clearly that Vandermonde had priority over Lagrange for this remarkable idea, which eventually led to the study of group theory.

Paolo Ruffini was the first person to develop the theory of permutation groups, and like his predecessors, also in the context of solving algebraic equations. His goal was to establish the impossibility of an algebraic solution to a general algebraic equation of degree greater than four. En route to this goal he introduced the notion of the order of an element of a group, conjugacy, the cycle decomposition of elements of permutation groups and the notions of primitive and imprimitive and proved some important theorems relating these concepts, such as

if G is a subgroup of S_5 whose order is divisible by 5 then G contains an element of order 5.

Note, however, that he got by without formalizing the concept of a group, or even of a permutation group. The next step was taken by Évariste Galois in 1832, although his work remained unpublished until 1846, when he considered for the first time what is now called the *closure property* of a group of permutations, which he expressed as

if in such a group one has the substitutions S and T then one has the substitution ST.

The theory of permutation groups received further far-reaching development in the hands of Augustin Cauchy and Camille Jordan, both through introduction of new concepts and, primarily, a great wealth of results about special classes of permutation groups and even some general theorems. Among other things, Jordan defined a notion of isomorphism, still in the context of permutation groups and, incidentally, it was he who put the term *group* in wide use.

The abstract notion of a group appeared for the first time in Arthur Cayley›s papers in 1854. Cayley realized that a group need not be a permutation group (or even *finite*), and may instead consist of matrices, whose algebraic properties, such as multiplication and inverses, he systematically investigated in succeeding years. Much later Cayley would revisit the question whether abstract groups were more general than permutation groups, and establish that, in fact, any group is isomorphic to a group of permutations.

Modern Algebra

The end of the 19th and the beginning of the 20th century saw a tremendous shift in the methodology of mathematics. Abstract algebra emerged around the start of the 20th century, under the name *modern algebra*. Its study was part of the drive for more intellectual rigor in mathematics. Initially, the assumptions in classical algebra, on which the whole of mathematics (and major parts of the natural sciences) depend, took the form of axiomatic systems. No longer satisfied with establishing properties of concrete objects, mathematicians started to turn their attention to general theory. Formal definitions of certain algebraic structures began to emerge in the 19th century. For example, results about various groups of permutations came to be seen as instances of general theorems that concern a general notion of an *abstract group*. Questions of structure and classification of various mathematical objects came to forefront.

These processes were occurring throughout all of mathematics, but became especially pronounced in algebra. Formal definition through primitive operations and axioms were proposed for many basic algebraic structures, such as groups, rings, and fields. Hence such things as group theory and ring theory took their places in pure mathematics. The algebraic investigations of general fields by Ernst Steinitz and of commutative and then general rings by David Hilbert, Emil Artin and Emmy Noether, building up on the work of Ernst Kummer, Leopold Kronecker and Richard Dedekind, who had considered ideals in commutative rings, and of Georg Frobenius and Issai Schur, concerning representation theory of groups, came to define abstract algebra. These developments of the last quarter of the 19th century and the first quarter of 20th century were systematically exposed in Bartel van der Waerden›s *Moderne algebra*, the two-volume monograph published in 1930–1931 that forever changed for the mathematical world the meaning of the word *algebra* from *the theory of equations* to the *theory of algebraic structures*.

BASIC CONCEPTS

By abstracting away various amounts of detail, mathematicians have defined various algebraic structures that are used in many areas of mathematics. For instance, almost all systems studied are sets, to which the theorems of set theory apply. Those sets that have a certain binary operation defined on them form magmas, to which the concepts concerning magmas, as well those concerning sets, apply. We can add additional constraints on the algebraic structure, such as associativity (to form semigroups); identity, and inverses (to form groups); and other more complex structures. With additional structure, more theorems could be proved, but the generality is reduced. The «hierarchy» of algebraic objects (in terms of generality) creates a hierarchy of the corresponding theories: for instance, the theorems of group theory may be used when studying rings (algebraic objects that have two binary operations with certain axioms) since a ring is a group over one of its operations. In general there is a balance between the amount of generality and the richness of the theory: more general structures have usually fewer nontrivial theorems and fewer applications.

Examples of algebraic structures with a single binary operation are:

- Magmas
- Quasigroups
- Monoids
- Semigroups
- Groups

Examples involving several operations include:

- Rings
- Fields
- Modules
- Vector spaces
- Algebras over fields
- Associative algebras
- Lie algebras
- Lattices
- Boolean algebras

ALGEBRAIC STRUCTURE

In mathematics, and more specifically in abstract algebra, an **algebraic structure** is a set (called **carrier set** or **underlying set**) with one or more finitary operations defined on it that satisfies a list of axioms.

Examples of algebraic structures include groups, rings, fields, and lattices. More complex structures can be defined by introducing multiple operations, different underlying sets, or by altering the defining axioms. Examples of more complex algebraic structures include vector spaces, modules, and algebras.

The properties of specific algebraic structures are studied in abstract algebra. The general theory of algebraic structures has been formalized in universal algebra. The language of category theory is used to express and study relationships between different classes of algebraic and non-algebraic objects. This because it is sometimes possible to find strong connections between some classes of objects, sometimes of different kinds. For example, Galois theory establishes a connection between certain fields and groups: two algebraic structures of different kinds.

Addition and multiplication on numbers are the prototypical example of an operation that combines two elements of a set to produce a third. These operations obey several algebraic laws. For example, $a+(b+c)=(a+b)+c$ and $a(bc)=(ab)c$, both examples of the *associative law*. Also $a+b=b+a$, and $ab=ba$, the *commutative law*. Many systems studied by mathematicians have operations that obey some, but not necessarily all, of the laws of ordinary arithmetic. For example, rotations of objects in three-dimensional space can be combined by performing the first rotation and then applying the second rotation to the object in its new orientation. This operation on rotations obeys the associative law, but can fail the commutative law.

Mathematicians give names to sets with one or more operations that obey a particular collection of laws, and study them in the abstract as algebraic structures. When a new problem can be shown to follow the laws of one of these algebraic structures, all the work that has been done on that category in the past can be applied to the new problem.

In full generality, algebraic structures may involve an arbitrary number of sets and operations that can combine more than two elements (higher arity), but this article focuses on binary operations on one or two sets. The examples here are by no means a complete list, but they are meant to be a representative list and include the most common structures. Longer lists of algebraic structures may be found in the external links and within *Category:Algebraic structures*. Structures are listed in approximate order of increasing complexity.

Examples

One Set with Operations

Simple structures: no binary operation:

- Set: a degenerate algebraic structure S having no operations.
- Pointed set: S has one or more distinguished elements, often 0, 1, or both.
- Unary system: S and a single unary operation over S.
- Pointed unary system: a unary system with S a pointed set.

Group-like structures: **one** binary operation. The binary operation can be indicated by any symbol, or with no symbol (juxtaposition) as is done for ordinary multiplication of real numbers.

- Magma or groupoid: S and a single binary operation over S.
- Semigroup: an associative magma.

- Monoid: a semigroup with identity.
- Group: a monoid with a unary operation (inverse), giving rise to inverse elements.
- Abelian group: a group whose binary operation is commutative.
- Semilattice: a semigroup whose operation is idempotent and commutative. The binary operation can be called either meet or join.
- Quasigroup: a magma obeying the latin square property. A quasigroup may also be represented using three binary operations.
- Loop: a quasigroup with identity.

Ring-like structures or *Ringoids*: **two** binary operations, often called addition and multiplication, with multiplication distributing over addition.

- Semiring: a ringoid such that S is a monoid under each operation. Addition is typically assumed to be commutative and associative, and the monoid product is assumed to distribute over the addition on both sides, and the additive identity satisfies $0\ x = 0$ for all x.
- Near-ring: a semiring whose additive monoid is a (not necessarily abelian) group.
- Ring: a semiring whose additive monoid is an abelian group.
- Lie ring: a ringoid whose additive monoid is an abelian group, but whose multiplicative operation satisfies the Jacobi identity rather than associativity.
- Boolean ring: a commutative ring with idempotent multiplication operation.
- Field: a commutative ring which contains a multiplicative inverse for every nonzero element
- Kleene algebras: a semiring with idempotent addition and a unary operation, the Kleene star, satisfying additional properties.
- *-algebra: a ring with an additional unary operation (*) satisfying additional properties.

Lattice structures: **two** or more binary operations, including operations called meet and join, connected by the absorption law.

- Complete lattice: a lattice in which arbitrary meet and joins exist.
- Bounded lattice: a lattice with a greatest element and least element.
- Complemented lattice: a bounded lattice with a unary operation, complementation, denoted by postfix $^{\perp}$. The join of an element with its complement is the greatest element, and the meet of the two elements is the least element.
- Modular lattice: a lattice whose elements satisfy the additional *modular identity*.
- Distributive lattice: a lattice in which each of meet and join distributes over the other. Distributive lattices are modular, but the converse does not hold.
- Boolean algebra: a complemented distributive lattice. Either of meet or join can be defined in terms of the other and complementation. This can be shown to be equivalent with the ring-like structure of the same name above.

- Heyting algebra: a bounded distributive lattice with an added binary operation, relative pseudo-complement, denoted by infix →, and governed by the axioms $x \to x = 1$, $x\ (x \to y) = x\ y$, $y\ (x \to y) = y$, $x \to (y\ z) = (x \to y)\ (x \to z)$.

Arithmetics: **two** binary operations, addition and multiplication. S is an infinite set. Arithmetics are pointed unary systems, whose unary operation is injective successor, and with distinguished element 0.

- Robinson arithmetic. Addition and multiplication are recursively defined by means of successor. 0 is the identity element for addition, and annihilates multiplication. Robinson arithmetic is listed here even though it is a variety, because of its closeness to Peano arithmetic.
- Peano arithmetic. Robinson arithmetic with an axiom schema of induction. Most ring and field axioms bearing on the properties of addition and multiplication are theorems of Peano arithmetic or of proper extensions thereof.

Two Sets with Operations

Module-like structures: composite systems involving two sets and employing at least two binary operations.

- Group with operators: a group G with a set Ω and a binary operation $\Omega \times G \to G$ satisfying certain axioms.
- Module: an Abelian group M and a ring R acting as operators on M. The members of R are sometimes called scalars, and the binary operation of *scalar multiplication* is a function $R \times M \to M$, which satisfies several axioms. Counting the ring operations these systems have at least three operations.
- Vector space: a module where the ring R is a division ring or field.
- Graded vector space: a vector space with a direct sum decomposition breaking the space into "grades".
- Quadratic space: a vector space V over a field F with a function from V into F satisfying certain properties. Every quadratic space is also an inner product space.

Algebra-like structures: composite system defined over two sets, a ring R and a R module M equipped with an operation called multiplication. This can be viewed as a system with five binary operations: two operations on R, two on M and one involving both R and M.

- Algebra over a ring (also *R-algebra*): a module over a commutative ring R, which also carries a multiplication operation that is compatible with the module structure. This includes distributivity over addition and linearity with respect to multiplication by elements of R. The theory of an algebra over a field is especially well developed.
- Associative algebra: an algebra over a ring such that the multiplication is associative.

- Nonassociative algebra: a module over a commutative ring, equipped with a ring multiplication operation that is not necessarily associative. Often associativity is replaced with a different identity, such as alternation, the Jacobi identity, or the Jordan identity.
- Coalgebra: a vector space with a «comultiplication» defined dually to that of associative algebras.
- Lie algebra: a special type of nonassociative algebra whose product satisfies the Jacobi identity.
- Lie coalgebra: a vector space with a «comultiplication» defined dually to that of Lie algebras.
- Graded algebra: a graded vector space with an algebra structure compatible with the grading. The idea is that if the grades of two elements *a* and *b* are known, then the grade of *ab* is known, and so the location of the product *ab* is determined in the decomposition.
- Inner product space: an *F* vector space *V* with a bilinear binary operation from $V \times V \to F$.

Four or more binary operations:

- Bialgebra: an associative algebra with a compatible coalgebra structure.
- Lie bialgebra: a Lie algebra with a compatible bialgebra structure.
- Hopf algebra: a bialgebra with a connection axiom (antipode).
- Clifford algebra: a graded associative algebra equipped with an exterior product from which may be derived several possible inner products. Exterior algebras and geometric algebras are special cases of this construction.

Hybrid Structures

Algebraic structures can also coexist with added structure of non-algebraic nature, such as partial order or a topology. The added structure must be compatible, in some sense, with the algebraic structure.

- Topological group: a group with a topology compatible with the group operation.
- Lie group: a topological group with a compatible smooth manifold structure.
- Ordered groups, ordered rings and ordered fields: each type of structure with a compatible partial order.
- Archimedean group: a linearly ordered group for which the Archimedean property holds.
- Topological vector space: a vector space whose *M* has a compatible topology.
- Normed vector space: a vector space with a compatible norm. If such a space is complete (as a metric space) then it is called a Banach space.
- Hilbert space: an inner product space over the real or complex numbers whose inner product gives rise to a Banach space structure.
- Vertex operator algebra

- Von Neumann algebra: a *-algebra of operators on a Hilbert space equipped with the weak operator topology.

Universal Algebra

Algebraic structures are defined through different configurations of axioms. Universal algebra abstractly studies such objects. One major dichotomy is between structures that are axiomatized entirely by *identities* and structures that are not. If all axioms defining a class of algebras are identities, then the class of objects is a variety (not to be confused with algebraic variety in the sense of algebraic geometry).

Identities are equations formulated using only the operations the structure allows, and variables that are tacitly universally quantified over the relevant universe. Identities contain no connectives, existentially quantified variables, or relations of any kind other than the allowed operations. The study of varieties is an important part of universal algebra. An algebraic structure in a variety may be understood as the quotient algebra of term algebra (also called "absolutely free algebra») divided by the equivalence relations generated by a set of identities. So, a collection of functions with given signatures generate a free algebra, the term algebra T. Given a set of equational identities (the axioms), one may consider their symmetric, transitive closure E. The quotient algebra T/E is then the algebraic structure or variety. Thus, for example, groups have a signature containing two operators: the multiplication operator m, taking two arguments, and the inverse operator i, taking one argument, and the identity element e, a constant, which may be considered an operator that takes zero arguments. Given a (countable) set of variables x, y, z, etc. the term algebra is the collection of all possible terms involving m, i, e and the variables; so for example, $m(i(x), m(x,m(y,e)))$ would be an element of the term algebra. One of the axioms defining a group is the identity $m(x, i(x)) = e$; another is $m(x,e) = x$. The axioms can be represented as trees. These equations induce equivalence classes on the free algebra; the quotient algebra then has the algebraic structure of a group.

Several non-variety structures fail to be varieties, because either:

- It is necessary that $0 \neq 1$, 0 being the additive identity element and 1 being a multiplicative identity element, but this is a nonidentity;
- Structures such as fields have some axioms that hold only for nonzero members of S. For an algebraic structure to be a variety, its operations must be defined for *all* members of S; there can be no partial operations.

Structures whose axioms unavoidably include nonidentities are among the most important ones in mathematics, e.g., fields and division rings. Although structures with nonidentities retain an undoubted algebraic flavor, they suffer from defects varieties do not have. For example, the product of two fields is not a field.

APPLICATIONS

Because of its generality, abstract algebra is used in many fields of mathematics and science. For instance, algebraic topology uses algebraic objects to study topologies. The Poincaré conjecture, proved in 2003, asserts that the fundamental group of a manifold, which encodes information about connectedness, can be used to determine whether a manifold is a sphere or not. Algebraic number theory studies various number rings that generalize the set of integers. Using tools of algebraic number theory, Andrew Wiles proved Fermat›s Last Theorem.

In physics, groups are used to represent symmetry operations, and the usage of group theory could simplify differential equations. In gauge theory, the requirement of local symmetry can be used to deduce the equations describing a system. The groups that describe those symmetries are Lie groups, and the study of Lie groups and Lie algebras reveals much about the physical system; for instance, the number of force carriers in a theory is equal to dimension of the Lie algebra, and these bosons interact with the force they mediate if the Lie algebra is nonabelian.

REFERENCES

1. Allenby, R.B.J.T. (1991), Rings, Fields and Groups, Butterworth-Heinemann, ISBN 978-0-340-54440-2
2. Artin, Michael (1991), Algebra, Prentice Hall, ISBN 978-0-89871-510-1
3. Burris, Stanley N.; Sankappanavar, H. P. (1999) [1981], A Course in Universal Algebra
4. Gilbert, Jimmie; Gilbert, Linda (2005), Elements of Modern Algebra, Thomson Brooks/Cole, ISBN 978-0-534-40264-8
5. Lang, Serge (2002), Algebra, Graduate Texts in Mathematics, 211 (Revised third ed.), New York: Springer-Verlag, ISBN 978-0-387-95385-4, MR1878556
6. Sethuraman, B. A. (1996), Rings, Fields, Vector Spaces, and Group Theory: An Introduction to Abstract Algebra via Geometric Constructibility, Berlin, New York: Springer-Verlag, ISBN 978-0-387-94848-5
7. Whitehead, C. (2002), Guide to Abstract Algebra (2nd ed.), Houndmills: Palgrave, ISBN 978-0-333-79447-0
8. W. Keith Nicholson (2012) Introduction to Abstract Algebra, 4th edition, John Wiley & Sons ISBN 978-1-118-13535-8 .
9. John R. Durbin (1992) Modern Algebra : an introduction, John Wiley & Sons
10. Mac Lane, Saunders; Birkhoff, Garrett (1999), Algebra (2nd ed.), AMS Chelsea, ISBN 978-0-8218-1646-2
11. Michel, Anthony N.; Herget, Charles J. (1993), Applied Algebra and Functional Analysis, New York: Dover Publications, ISBN 978-0-486-67598-5
12. Burris, Stanley N.; Sankappanavar, H. P. (1981), A Course in Universal Algebra, Berlin, New York: Springer-Verlag, ISBN 978-3-540-90578-3

CHAPTER 7

Discrete Geometry

Chapter Contents

- **Introduction**
- **Topics in Discrete Geometry**

Discrete geometry is concerned with properties of finitely generated geometric objects such as polytopes and polyhedra, triangulations and polyhedral complexes, configurations of lines and, more generally, hyperplanes in Euclidean and other spaces, the theory of rigid and flexible frameworks, tilings and packings. Many problems in discrete geometry arise from questions in computational geometry related to algorithms for analyzing discrete geometric structures.

INTRODUCTION

Discrete geometry and combinatorial geometry are branches of geometry that study combinatorial properties and constructive methods of discrete geometric objects. Most questions in discrete geometry involve finite or discrete sets of basic geometric objects, such as points, lines, planes, circles, spheres, polygons, and so forth. The subject focuses on the combinatorial properties of these objects, such as how they intersect one another, or how they may be arranged to cover a larger object.

Discrete geometry has large overlap with convex geometry and computational geometry, and is closely related to subjects such as finite geometry, combinatorial optimization, digital geometry, discrete differential geometry, geometric graph theory, toric geometry, and combinatorial topology.

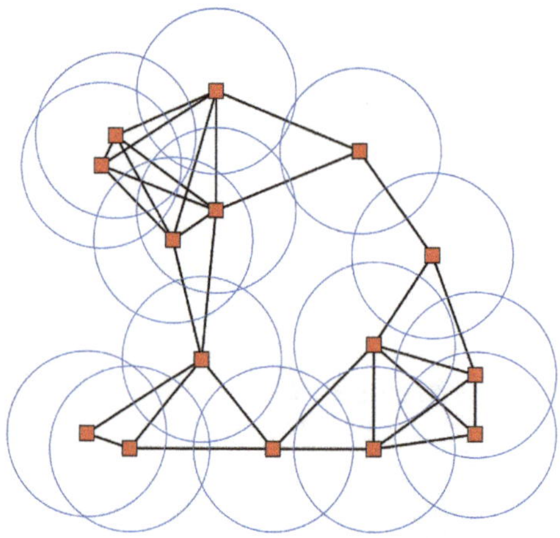

A collection of circles and the corresponding unit disk graph

Although polyhedra and tessellations had been studied for many years by people such as Kepler and Cauchy, modern discrete geometry has its origins in the late 19th century. Early topics studied were: the density of circle packings by Thue, projective configurations by Reye and Steinitz, the geometry of numbers by Minkowski, and map colourings by Tait, Heawood, and Hadwiger.

László Fejes Tóth, H.S.M. Coxeter and Paul Erdős, laid the foundations of *discrete geometry*.

TOPICS IN DISCRETE GEOMETRY

Polyhedra and Polytopes

A **polytope** is a geometric object with flat sides, which exists in any general number of dimensions. A polygon is a polytope in two dimensions, a polyhedron in three dimensions, and so on in higher dimensions (such as a 4-polytope in four dimensions). Some theories further generalize the idea to include such objects as unbounded polytopes (apeirotopes and tessellations), and abstract polytopes.

The following are some of the aspects of polytopes studied in discrete geometry:

- Polyhedral combinatorics
- Lattice polytopes
- Ehrhart polynomials

- Pick›s theorem
- Hirsch conjecture

Packings, Coverings and Tilings

Packings, coverings, and tilings are all ways of arranging uniform objects (typically circles, spheres, or tiles) in a regular way on a surface or manifold.

A **sphere packing** is an arrangement of non-overlapping spheres within a containing space. The spheres considered are usually all of identical size, and the space is usually three-dimensional Euclidean space. However, sphere packing problems can be generalised to consider unequal spheres, *n*-dimensional Euclidean space (where the problem becomes circle packing in two dimensions, or hypersphere packing in higher dimensions) or to non-Euclidean spaces such as hyperbolic space.

A **tessellation** of a flat surface is the tiling of a plane using one or more geometric shapes, called tiles, with no overlaps and no gaps. In mathematics, tessellations can be generalized to higher dimensions.

Specific topics in this area include:

- Circle packings
- Sphere packings
- Kepler conjecture
- Quasicrystals
- Aperiodic tilings
- Periodic Graphs (Geometry)
- Finite subdivision rules

Structural Rigidity and Flexibility

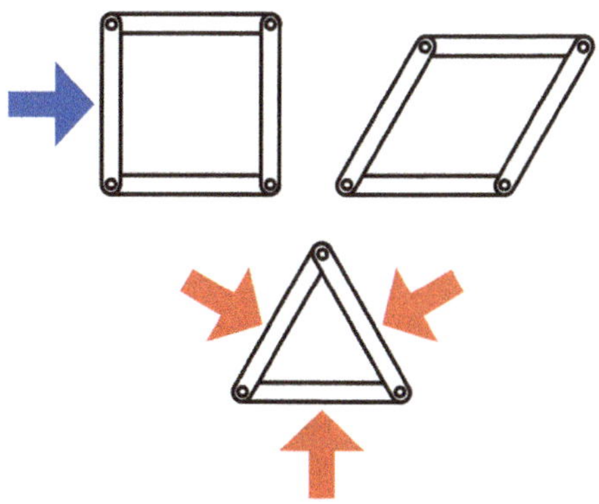

Graphs are drawn as rods connected by rotating hinges. The cycle graph C_4 drawn as a square can be tilted over by the blue force into a parallelogram, so it is a flexible graph. K_3, drawn as a triangle, cannot be altered by any force that is applied to it, so it is a rigid graph.

Structural rigidity is a combinatorial theory for predicting the flexibility of ensembles formed by rigid bodies connected by flexible linkages or hinges.

Topics in this area include:

- Cauchy›s theorem
- Flexible polyhedra

Incidence Structures

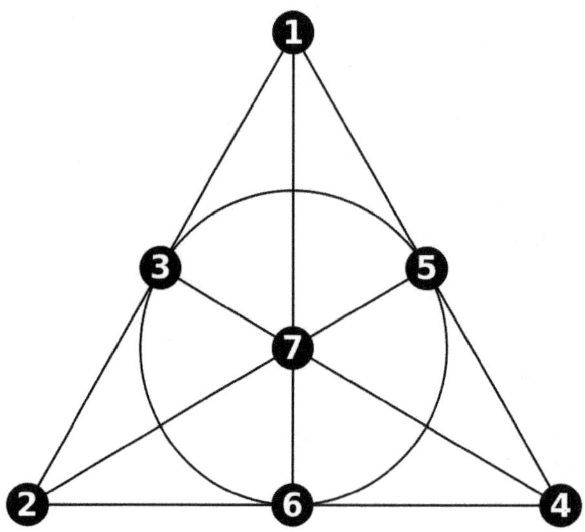

Seven points are elements of seven lines in the Fano plane, an example of an incidence structure.

Incidence structures generalize planes (such as affine, projective, and Möbius planes) as can be seen from their axiomatic definitions. Incidence structures also generalize the higher-dimensional analogs and the finite structures are sometimes called finite geometries.

Formally, an **incidence structure** is a triple

$$C = (P, L, I).$$

where P is a set of "points", L is a set of "lines" and $I \subseteq P \times L$ is the incidence relation. The elements of I are called **flags.** If

$$(p, l) \in I,$$

we say that point p "lies on" line l.

Topics in this area include:

- Configurations
- Line arrangements
- Hyperplane arrangements
- Buildings

Oriented Matroids

An **oriented matroid** is a mathematical structure that abstracts the properties of directed graphs and of arrangements of vectors in a vector space over an ordered field (particularly for partially ordered vector spaces). In comparison, an ordinary (i.e., non-oriented) matroid abstracts the dependence properties that are common both to graphs, which are not necessarily *directed*, and to arrangements of vectors over fields, which are not necessarily *ordered*.

Geometric Graph Theory

A **geometric graph** is a graph in which the vertices or edges are associated with geometric objects. Examples include Euclidean graphs, the 1-skeleton of a polyhedron or polytope, intersection graphs, and visibility graphs.

Topics in this area include:

- Graph drawing
- Polyhedral graphs
- Voronoi diagrams and Delaunay triangulations

Simplicial Complexes

A **simplicial complex** is a topological space of a certain kind, constructed by "gluing together" points, line segments, triangles, and their n-dimensional counterparts (see illustration). Simplicial complexes should not be confused with the more abstract notion of a simplicial set appearing in modern simplicial homotopy theory. The purely combinatorial counterpart to a simplicial complex is an abstract simplicial complex.

Topological Combinatorics

The discipline of combinatorial topology used combinatorial concepts in topology and in the early 20th century this turned into the field of algebraic topology.

In 1978 the situation was reversed – methods from algebraic topology were used to solve a problem in combinatorics – when László Lovász proved the Kneser conjecture, thus beginning the new study of **topological combinatorics**. Lovász's proof used the Borsuk-Ulam theorem and this theorem retains a prominent role in this new field. This theorem has many equivalent versions and analogs and has been used in the study of fair division problems.

Topics in this are include:

- Sperner›s lemma
- Regular maps

Lattices and Discrete Groups

A **discrete group** is a group *G* equipped with the discrete topology. With this topology, *G* becomes a topological group. A **discrete subgroup** of a topological group *G* is a subgroup *H* whose relative topology is the discrete one. For example, the integers, **Z**, form a discrete subgroup of the reals, **R** (with the standard metric topology), but the rational numbers, **Q**, do not.

A **lattice** in a locally compact topological group is a discrete subgroup with the property that the quotient space has finite invariant measure. In the special case of subgroups of $\mathbf{R}^n$, this amounts to the usual geometric notion of a lattice, and both the algebraic structure of lattices and the geometry of the totality of all lattices are relatively well understood. Deep results of Borel, Harish-Chandra, Mostow, Tamagawa, M. S. Raghunathan, Margulis, Zimmer obtained from the 1950s through the 1970s provided examples and generalized much of the theory to the setting of nilpotent Lie groups and semisimple algebraic groups over a local field. In the 1990s, Bass and Lubotzky initiated the study of *tree lattices*, which remains an active research area.

Topics in this area include:

- Reflection groups
- Triangle groups

Digital Geometry

Digital geometry deals with discrete sets (usually discrete point sets) considered to be digitized models or images of objects of the 2D or 3D Euclidean space.

Simply put, **digitizing** is replacing an object by a discrete set of its points. The images we see on the TV screen, the raster display of a computer, or in newspapers are in fact digital images.

Its main application areas are computer graphics and image analysis. See Li Chen, Digital and discrete geometry: Theory and Algorithms, Springer, 2014. (http://www.springer.com/us/book/9783319120980)

Discrete Differential Geometry

Discrete differential geometry is the study of discrete counterparts of notions in differential geometry. Instead of smooth curves and surfaces, there are polygons, meshes, and simplicial complexes. It is used in the study of computer graphics and topological combinatorics.

Topics in this area include:

- Discrete Laplace operator

- Discrete exterior calculus
- Discrete Morse theory
- Topological combinatorics
- Spectral shape analysis
- Abstract differential geometry
- Analysis on fractals

REFERENCES

1. Bezdek, András, (2003). Discrete geometry: in honor of W. Kuperberg's 60th birthday. New York, N.Y: Marcel Dekker. ISBN 0-8247-0968-3.
2. Bezdek, Károly (2010). Classical Topics in Discrete Geometry. New York, N.Y: Springer. ISBN 978-1-4419-0599-4.
3. Bezdek, Károly (2013). Lectures on Sphere Arrangements - the Discrete Geometric Side. New York, N.Y: Springer. ISBN 978-1-4614-8117-1.
4. Bezdek, Károly; Deza, Antoine; Ye, Yinyu (2013). Discrete Geometry and Optimization. New York, N.Y: Springer. ISBN 978-3-319-00200-2.
5. Brass, Peter; Moser, William; Pach, János (2005). Research problems in discrete geometry. Berlin: Springer. ISBN 0-387-23815-8.
6. Pach, János; Agarwal, Pankaj K. (1995). Combinatorial geometry. New York: Wiley-Interscience. ISBN 0-471-58890-3.
7. Goodman, Jacob E. and O'Rourke, Joseph (2004). Handbook of Discrete and Computational Geometry, Second Edition. Boca Raton: Chapman & Hall/CRC. ISBN 1-58488-301-4.
8. Gruber, Peter M. (2007). Convex and Discrete Geometry. Berlin: Springer. ISBN 3-540-71132-5.
9. Matoušek, Jiří (2002). Lectures on discrete geometry. Berlin: Springer. ISBN 0-387-95374-4.
10. Vladimir Boltyanski, Horst Martini, Petru S. Soltan, (1997). Excursions into Combinatorial Geometry. Springer. ISBN 3-540-61341-2.

CHAPTER 8

Game Theory

Chapter Contents

- **Introduction**

Game theory is "the study of mathematical models of conflict and cooperation between intelligent rational decision-makers." Game theory is mainly used in economics, political science, and psychology, as well as logic, computer science, biology and poker. Originally, it addressed zero-sum games, in which one person›s gains result in losses for the other participants. Today, game theory applies to a wide range of behavioral relations, and is now an umbrella term for the science of logical decision making in humans, animals, and computers.

INTRODUCTION

Modern game theory began with the idea regarding the existence of mixed-strategy equilibria in two-person zero-sum games and its proof by John von Neumann. Von Neumann's original proof used Brouwer fixed-point theorem on continuous mappings into compact convex sets, which became a standard method in game theory and mathematical economics. His paper was followed by the 1944 book *Theory of Games and Economic Behavior*, co-written with Oskar Morgenstern, which considered cooperative games of several players. The second edition of this book provided an axiomatic theory of expected utility, which allowed mathematical statisticians and economists to treat decision-making under uncertainty.

This theory was developed extensively in the 1950s by many scholars. Game theory was later explicitly applied to biology in the 1970s, although similar developments go back at least as far as the 1930s. Game theory has been widely recognized as an important tool in many fields. With the Nobel Memorial Prize in Economic Sciences going to game theorist Jean Tirole in 2014, eleven game-theorists have now won the economics Nobel Prize. John Maynard Smith was awarded the Crafoord Prize for his application of game theory to biology.

John von Neumann

Early discussions of examples of two-person games occurred long before the rise of modern, mathematical game theory. The first known discussion of game theory occurred in a letter written by Charles Waldegrave, an active Jacobite, and uncle to James Waldegrave, a British diplomat, in 1713. In this letter, Waldegrave provides a minimax mixed strategy solution to a two-person version of the card game le Her, and the problem is now known as Waldegrave problem. James Madison made what we now recognize as a game-theoretic analysis of the ways states can be expected to behave under different systems of taxation. In his 1838 *Recherches sur les principes mathématiques de la théorie des richesses* (*Researches into the Mathematical Principles of the Theory of Wealth*), Antoine Augustin Cournot considered a duopoly and presents a solution that is a restricted version of the Nash equilibrium.

In 1913 Ernst Zermelo published *Über eine Anwendung der Mengenlehre auf die Theorie des Schachspiels*. It proved that the optimal chess strategy is strictly determined. This paved the way for more general theorems.

The Danish mathematician Zeuthen proved that the mathematical model had a winning strategy by using Brouwer's fixed point theorem. In his 1938 book *Applications aux Jeux de Hasard* and earlier notes, Émile Borel proved a minimax theorem for two-person zero-sum matrix games only when the pay-off matrix was symmetric. Borel conjectured that non-existence of mixed-strategy equilibria in two-person zero-sum games would occur, a conjecture that was proved false.

Game theory did not really exist as a unique field until John von Neumann published a paper in 1928. Von Neumann's original proof used Brouwer›s fixed-point theorem on continuous mappings into compact convex sets, which became a standard method in game theory and mathematical economics. His paper was followed by his 1944 book *Theory of Games and Economic Behavior* co-authored with Oskar Morgenstern. The second edition of this book provided an axiomatic theory of utility, which reincarnated Daniel Bernoulli›s old theory of utility (of the money) as an independent discipline. Von Neumann›s work in game theory culminated in this 1944 book. This foundational work contains the method for finding mutually consistent solutions for two-person zero-sum games. During the following time period, work on game theory was primarily focused on cooperative game theory, which analyzes optimal strategies for groups of individuals, presuming that they can enforce agreements between them about proper strategies.

In 1950, the first mathematical discussion of the prisoner›s dilemma appeared, and an experiment was undertaken by notable mathematicians Merrill M. Flood and Melvin Dresher, as part of the RAND Corporation›s investigations into game theory. RAND pursued the studies because of possible applications to global nuclear strategy. Around this same time, John Nash developed a criterion for mutual consistency of players' strategies, known as Nash equilibrium, applicable to a wider variety of games than the criterion proposed by von Neumann and Morgenstern. This equilibrium is sufficiently general to allow for the analysis of non-cooperative games in addition to cooperative ones.

Game theory experienced a flurry of activity in the 1950s, during which time the concepts of the core, the extensive form game, fictitious play, repeated games, and the Shapley value were developed. In addition, the first applications of game theory to philosophy and political science occurred during this time.

Prize-winning achievements

In 1965, Reinhard Selten introduced his solution concept of subgame perfect equilibria, which further refined the Nash equilibrium (later he would introduce trembling hand perfection as well). In 1967, John Harsanyi developed the concepts of complete information and Bayesian games. Nash, Selten and Harsanyi became Economics Nobel Laureates in 1994 for their contributions to economic game theory.

In the 1970s, game theory was extensively applied in biology, largely as a result of the work of John Maynard Smith and his evolutionarily stable strategy. In addition, the concepts of correlated equilibrium, trembling hand perfection, and common knowledge were introduced and analyzed.

In 2005, game theorists Thomas Schelling and Robert Aumann followed Nash, Selten and Harsanyi as Nobel Laureates. Schelling worked on dynamic models, early examples of evolutionary game theory. Aumann contributed more to the equilibrium school, introducing an equilibrium coarsening, correlated equilibrium, and developing an extensive formal analysis of the assumption of common knowledge and of its consequences.

In 2007, Leonid Hurwicz, together with Eric Maskin and Roger Myerson, was awarded the Nobel Prize in Economics for having laid the foundations of mechanism design theory." Myerson's contributions include the notion of proper equilibrium, and an important graduate text: *Game Theory, Analysis of Conflict*. Hurwicz introduced and formalized the concept of incentive compatibility.

In 2012, Alvin E. Roth and Lloyd S. Shapley were awarded the Nobel Prize in Economics "for the theory of stable allocations and the practice of market design" and, in 2014, the Nobel went to game theorist Jean Tirole.

Game Types

Cooperative / Non-Cooperative

A game is *cooperative* if the players are able to form binding commitments externally enforced (e.g. through contract law). A game is *non-cooperative* if players cannot form alliances or if all agreements need to be self-enforcing (e.g. through credible threats).

Cooperative games are often analysed through the framework of *cooperative game theory*, which focuses on predicting which coalitions will form, the joint actions that groups take and the resulting collective payoffs. It is opposed to the traditional *non-cooperative game theory* which focuses on predicting individual players' actions and payoffs and analyzing Nash equilibria.

Cooperative game theory provides a high-level approach as it only describes the structure, strategies and payoffs of coalitions, whereas non-cooperative game theory also looks at how bargaining procedures will affect the distribution of payoffs within each coalition. As non-cooperative game theory is more general, cooperative games can be analyzed through the approach of non-cooperative game theory (the converse does not hold) provided that sufficient assumptions are made to encompass all the possible strategies available to players due to the possibility of external enforcement of cooperation. While it would thus be optimal to have all games expressed under a non-cooperative framework, in many instances insufficient information is available to accurately model the formal procedures available to the players during the strategic bargaining process, or the resulting model would be of too high complexity to offer a practical tool in the real world. In such cases, cooperative game theory provides a simplified approach that allows to analyze the game at large without having to make any assumption about bargaining powers.

Symmetric / Asymmetric

A symmetric game is a game where the payoffs for playing a particular strategy depend only on the other strategies employed, not on who is playing them. If the identities of the players can be changed without changing the payoff to the strategies, then a game is symmetric. Many of the commonly studied 2×2 games are symmetric. The standard representations of chicken, the prisoner's dilemma, and the stag hunt are all symmetric games. Some scholars would consider certain asymmetric games as examples of these games as well. However, the most common payoffs for each of these games are symmetric.

	E	F
E	1, 2	0, 0
F	0, 0	1, 2
An asymmetric game		

Most commonly studied asymmetric games are games where there are not identical strategy sets for both players. For instance, the ultimatum game and similarly the dictator game have different strategies for each player. It is possible, however, for a game to have identical strategies for both players, yet be asymmetric. For example, the game pictured to the right is asymmetric despite having identical strategy sets for both players.

Zero-sum / Non-zero-sum

Zero-sum games are a special case of constant-sum games, in which choices by players can neither increase nor decrease the available resources. In zero-sum games the total benefit to all players in the game, for every combination of strategies, always adds to zero (more informally, a player benefits only at the equal expense of others). Poker exemplifies a zero-sum game (ignoring the possibility of the house's cut), because one wins exactly the amount one's opponents lose. Other zero-sum games include matching pennies and most classical board games including Go and chess.

	A	B
A	–1, 1	3, –3
B	0, 0	–2, 2
A zero-sum game		

Many games studied by game theorists (including the famed prisoner's dilemma) are non-zero-sum games, because the outcome has net results greater or less than zero. Informally, in non-zero-sum games, a gain by one player does not necessarily correspond with a loss by another.

Constant-sum games correspond to activities like theft and gambling, but not to the fundamental economic situation in which there are potential gains from trade. It is possible to transform any game into a (possibly asymmetric) zero-sum game by adding a dummy player (often called the board) whose losses compensate the players› net winnings.

Simultaneous / Sequential

Simultaneous games are games where both players move simultaneously, or if they do not move simultaneously, the later players are unaware of the earlier players' actions (making them *effectively* simultaneous). Sequential games (or dynamic games) are games where later players have some knowledge about earlier actions. This need not be perfect information about every action of earlier players; it might be very little knowledge. For instance, a player may know that an earlier player did not perform one particular action, while he does not know which of the other available actions the first player actually performed.

The difference between simultaneous and sequential games is captured in the different representations discussed above. Often, normal form is used to represent simultaneous games, while extensive form is used to represent sequential ones. The transformation of extensive to normal form is one way, meaning that multiple extensive form games correspond to the same normal form. Consequently, notions of equilibrium for simultaneous games are insufficient for reasoning about sequential games; see subgame perfection.

In short, the differences between sequential and simultaneous games are as follows:

	Sequential	**Simultaneous**
Normally denoted by	Decision trees	Payoff matrices
Prior knowledge of opponent's move?	Yes	No
Time axis?	Yes	No
Also known as	Extensive-form game Extensive game	Strategy game Strategic game

Perfect Information and Imperfect Information

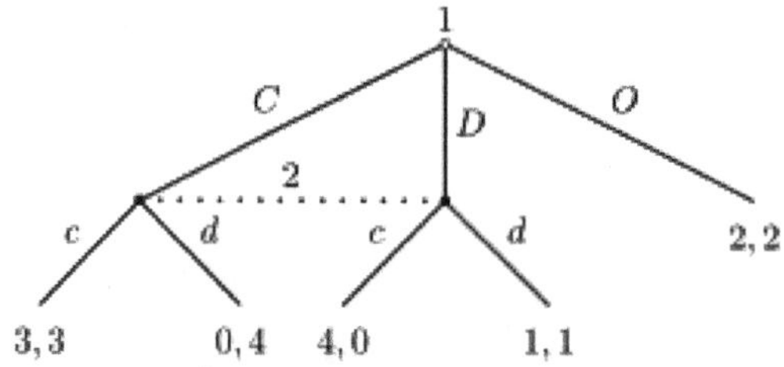

A game of imperfect information (the dotted line represents ignorance on the part of player 2, formally called an information set)

An important subset of sequential games consists of games of perfect information. A game is one of perfect information if, in extensive form, all players know the moves previously made by all other players. Simultaneous games can not be games of perfect information, because the conversion to extensive form converts simultaneous moves into a sequence of moves with earlier moves being unknown. Most games studied in game theory are imperfect-information games. Interesting examples of perfect-information games include the ultimatum game and centipede game. Recreational games of perfect information games include chess and checkers. Many card games are games of imperfect information, such as poker or contract bridge.

Perfect information is often confused with complete information, which is a similar concept. Complete information requires that every player know the strategies and payoffs available to the other players but not necessarily the actions taken. Games of incomplete information can be reduced, however, to games of imperfect information by introducing moves by nature.

Combinatorial Games

Games in which the difficulty of finding an optimal strategy stems from the multiplicity of possible moves are called combinatorial games. Examples include chess and go. Games that involve imperfect or incomplete information may also have a strong combinatorial character, for instance backgammon. There is no unified theory addressing combinatorial elements in games. There are, however, mathematical tools that can solve particular problems and answer general questions.

Games of perfect information have been studied in combinatorial game theory, which has developed novel representations, e.g. surreal numbers, as well as combinatorial and algebraic (and sometimes non-constructive) proof methods to solve games of certain types, including "loopy" games that may result in infinitely long sequences of moves. These methods address games with higher combinatorial complexity than those usually considered in traditional (or "economic") game theory. A typical game that has been solved this way is hex. A related field of study, drawing from computational complexity theory, is game complexity, which is concerned with estimating the computational difficulty of finding optimal strategies.

Research in artificial intelligence has addressed both perfect and imperfect (or incomplete) information games that have very complex combinatorial structures (like chess, go, or backgammon) for which no provable optimal strategies have been found. The practical solutions involve computational heuristics, like alpha-beta pruning or use of artificial neural networks trained by reinforcement learning, which make games more tractable in computing practice.

Infinitely Long Games

Games, as studied by economists and real-world game players, are generally finished in finitely many moves. Pure mathematicians are not so constrained, and set theorists in particular study games that last for infinitely many moves, with the winner (or other payoff) not known until *after* all those moves are completed.

The focus of attention is usually not so much on the best way to play such a game, but whether one player has a winning strategy. (It can be proven, using the axiom of choice, that there are games – even with perfect information and where the only outcomes are win or lose – for which *neither* player has a winning strategy.) The existence of such strategies, for cleverly designed games, has important consequences in descriptive set theory.

Discrete and Continuous Games

Much of game theory is concerned with finite, discrete games, that have a finite number of players, moves, events, outcomes, etc. Many concepts can be extended, however. Continuous games allow players to choose a strategy from a continuous strategy set. For instance, Cournot competition is typically modeled with players' strategies being any non-negative quantities, including fractional quantities.

Differential Games

Differential games such as the continuous pursuit and evasion game are continuous games where the evolution of the players' state variables is governed by differential equations. The problem of finding an optimal strategy in a differential game is closely related to the optimal control theory. In particular, there are two types of strategies: the open-loop strategies are found using the Pontryagin maximum principle while the closed-loop strategies are found using Bellman›s Dynamic Programming method.

A particular case of differential games are the games with a random time horizon. In such games, the terminal time is a random variable with a given probability distribution function. Therefore, the players maximize the mathematical expectation of the cost function. It was shown that the modified optimization problem can be reformulated as a discounted differential game over an infinite time interval.

Many-player and population games

Games with an arbitrary, but finite, number of players are often called n-person games. Evolutionary game theory considers games involving a population of decision makers, where the frequency with which a particular decision is made can change over time in response to the decisions made by all individuals in the population. In biology, this is intended to model (biological) evolution, where genetically programmed organisms pass along some of their strategy programming to their offspring. In economics, the same theory is intended to capture population changes because people play the game many times within their lifetime, and consciously (and perhaps rationally) switch strategies.

Stochastic outcomes (and relation to other fields)

Individual decision problems with stochastic outcomes are sometimes considered "one-player games". These situations are not considered game theoretical by some authors. They may be modeled using similar tools within the related disciplines of decision theory, operations research, and areas of artificial intelligence, particularly AI planning (with uncertainty) and multi-agent system. Although these fields may have different motivators, the mathematics involved are substantially the same, e.g. using Markov decision processes (MDP).

Stochastic outcomes can also be modeled in terms of game theory by adding a randomly acting player who makes chance moves (moves by nature). This player is not typically considered a third player in what is otherwise a two-player game, but merely serves to provide a roll of the dice where required by the game.

For some problems, different approaches to modeling stochastic outcomes may lead to different solutions. For example, the difference in approach between MDPs and the minimax solution is that the latter considers the worst-case over a set of adversarial moves, rather than reasoning in expectation about these moves given a fixed probability distribution. The minimax approach may be advantageous where stochastic models of uncertainty are not available, but may also be overestimating extremely unlikely (but costly) events, dramatically swaying the strategy in such scenarios if it is assumed that an adversary can force such an event to happen.(See Black swan theory for more discussion on this kind of modeling issue, particularly as it relates to predicting and limiting losses in investment banking.)

General models that include all elements of stochastic outcomes, adversaries, and partial or noisy observability (of moves by other players) have also been studied. The gold standard is considered to be partially observable stochastic game (POSG), but few realistic problems are computationally feasible in POSG representation.

Metagames

These are games the play of which is the development of the rules for another game, the target or subject game. Metagames seek to maximize the utility value of the rule set developed. The theory of metagames is related to mechanism design theory.

The term metagame analysis is also used to refer to a practical approach developed by Nigel Howard. whereby a situation is framed as a strategic game in which stakeholders try to realise their objectives by means of the options available to them. Subsequent developments have led to the formulation of confrontation analysis.

Pooling games

These are games prevailing over all forms of society. Pooling games are repeated plays with changing payoff table in general over an experienced path and their equilibrium strategies usually take a form of evolutionary social convention and economic convention. Pooling game theory emerges to formally recognize the interaction between optimal choice in one play and the emergence of forthcoming payoff table update path, identify the invariance existence and robustness, and predict variance over time. The theory is based upon topological transformation classification of payoff table update over time to predict variance and invariance, and is also within the jurisdiction of the computational law of reachable optimality for ordered system.

Representation of Games

List of games in game theory

The games studied in game theory are well-defined mathematical objects. To be fully defined, a game must specify the following elements: the *players* of the game, the *information* and *actions* available to each player at each decision point, and the *payoffs* for each outcome. (Eric Rasmusen refers to these four essential elements by the acronym

PAPI.) A game theorist typically uses these elements, along with a solution concept of their choosing, to deduce a set of equilibrium strategies for each player such that, when these strategies are employed, no player can profit by unilaterally deviating from their strategy. These equilibrium strategies determine an equilibrium to the game—a stable state in which either one outcome occurs or a set of outcomes occur with known probability.

Most cooperative games are presented in the characteristic function form, while the extensive and the normal forms are used to define noncooperative games.

Extensive form

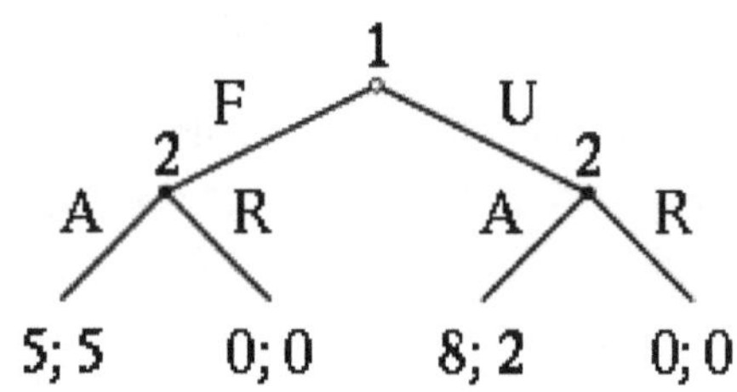

An extensive form game

The extensive form can be used to formalize games with a time sequencing of moves. Games here are played on trees (as pictured here). Here each vertex (or node) represents a point of choice for a player. The player is specified by a number listed by the vertex. The lines out of the vertex represent a possible action for that player. The payoffs are specified at the bottom of the tree. The extensive form can be viewed as a multi-player generalization of a decision tree. To solve any extensive form game, the backwards induction must be used. It involves working backwards up the game tree to determine what a rational player would do at the last vertex of the tree, what the player with the previous move would do given that the player with the last move is rational, and so on until the first vertex of the tree is reached.

The game pictured consists of two players. The way this particular game is structured (i.e., with sequential decision making and perfect information), *Player 1* "moves" first by choosing either *F* or *U* (letters are assigned arbitrarily for mathematical purposes). Next in the sequence, *Player 2*, who has now seen *Player 1*'s move, chooses to play either *A* or *R*. Once *Player 2* has made his/ her choice, the game is considered finished and each player gets their respective payoff. Suppose that *Player 1* chooses *U* and then *Player 2* chooses *A*: *Player 1* then gets a payoff of "eight" (which in real-world terms can be interpreted in many ways, the simplest of which is in terms of money but could mean things such as eight days of vacation or eight countries conquered or even eight more opportunities to play the same game against other players) and *Player 2* gets a payoff of "two".

The extensive form can also capture simultaneous-move games and games with imperfect information. To represent it, either a dotted line connects different vertices to represent them as being part of the same information set (i.e. the players do not know at which point they are), or a closed line is drawn around them. (See example in the imperfect information section.)

Normal form

The normal (or strategic form) game is usually represented by a matrix which shows the players, strategies, and payoffs (see the example to the right). More generally it can be represented by any function that associates a payoff for each player with every possible combination of actions. In the accompanying example there are two players; one chooses the row and the other chooses the column. Each player has two strategies, which are specified by the number of rows and the number of columns. The payoffs are provided in the interior. The first number is the payoff received by the row player (Player 1 in our example); the second is the payoff for the column player (Player 2 in our example). Suppose that Player 1 plays *Up* and that Player 2 plays *Left*. Then Player 1 gets a payoff of 4, and Player 2 gets 3.

	Player 2 chooses *Left*	Player 2 chooses *Right*
Player 1 chooses *Up*	**4, 3**	**−1, −1**
Player 1 chooses *Down*	**0, 0**	**3, 4**

Normal form or payoff matrix of a 2-player, 2-strategy game

When a game is presented in normal form, it is presumed that each player acts simultaneously or, at least, without knowing the actions of the other. If players have some information about the choices of other players, the game is usually presented in extensive form.

Every extensive-form game has an equivalent normal-form game, however the transformation to normal form may result in an exponential blowup in the size of the representation, making it computationally impractical.

Characteristic function form

In games that possess removable utility, separate rewards are not given; rather, the characteristic function decides the payoff of each unity. The idea is that the unity that is 'empty', so to speak, does not receive a reward at all.

The origin of this form is to be found in John von Neumann and Oskar Morgenstern's book; when looking at these instances, they guessed that when a union C appears, it works against the fraction $\left(\frac{N}{C}\right)$ as if two individuals were playing a normal game. The balanced payoff of C is a basic function. Although there are differing examples that help determine coalitional amounts from normal games, not all appear that in their function form can be derived from such.

Formally, a characteristic function is seen as: (N,v), where N represents the group of people and $v : 2^N \to \mathbf{R}$ is a normal utility.

Such characteristic functions have expanded to describe games where there is no removable utility.

General and Applied Uses

As a method of applied mathematics, game theory has been used to study a wide variety of human and animal behaviors. It was initially developed in economics to understand a large collection of economic behaviors, including behaviors of firms, markets, and consumers. The first use of game-theoretic analysis was by Antoine Augustin Cournot in 1838 with his solution of the Cournot duopoly. The use of game theory in the social sciences has expanded, and game theory has been applied to political, sociological, and psychological behaviors as well.

Although pre-twentieth century naturalists such as Charles Darwin made game-theoretic kinds of statements, the use of game-theoretic analysis in biology began with Ronald Fisher›s studies of animal behavior during the 1930s. This work predates the name game theory, but it shares many important features with this field. The developments in economics were later applied to biology largely by John Maynard Smith in his book *Evolution and the Theory of Games*.

In addition to being used to describe, predict, and explain behavior, game theory has also been used to develop theories of ethical or normative behavior and to prescribe such behavior. In economics and philosophy, scholars have applied game theory to help in the understanding of good or proper behavior. Game-theoretic arguments of this type can be found as far back as Plato.

Description and Modeling

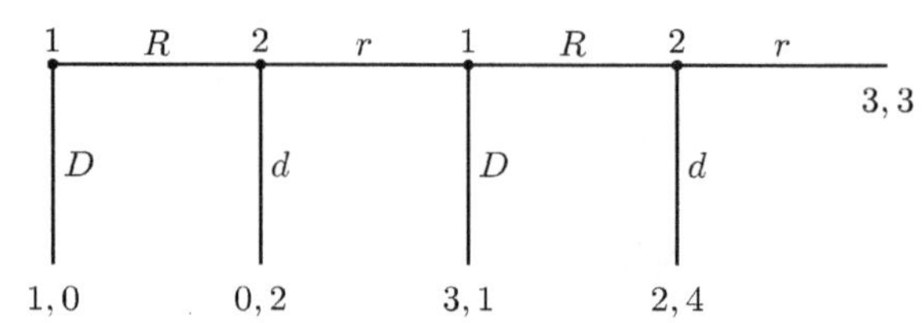

A four-stage centipede game

The primary use of game theory is to describe and model how human populations behave. Some scholars believe that by finding the equilibria of games they can predict how actual human populations will behave when confronted with situations analogous to the game being studied. This particular view of game theory has been criticized. It is argued that the assumptions made by game theorists are often violated when applied to real world situations. Game theorists usually assume players act rationally, but in practice, human behavior often deviates from this model. Game theorists respond by comparing their assumptions to those used in physics. Thus while their assumptions do not always hold, they can treat game theory as a reasonable scientific ideal akin to the models used by physicists. However, empirical work has shown that in some classic games, such as the centipede game, guess 2/3 of the average game, and the dictator game, people regularly do not play Nash equilibria. There is an ongoing debate regarding the importance of these experiments and whether the analysis of the experiments fully captures all aspects of the relevant situation.

Some game theorists, following the work of John Maynard Smith and George R. Price, have turned to evolutionary game theory in order to resolve these issues. These models presume either no rationality or bounded rationality on the part of players. Despite the name, evolutionary game theory does not necessarily presume natural selection in the biological sense. Evolutionary game theory includes both biological as well as cultural evolution and also models of individual learning (for example, fictitious play dynamics).

Prescriptive or Normative Analysis

	Cooperate	Defect
Cooperate	-1, -1	-10, 0
Defect	0, -10	-5, -5
The Prisoner's Dilemma		

Some scholars, like Leonard Savage, see game theory not as a predictive tool for the behavior of human beings, but as a suggestion for how people ought to behave. Since a strategy, corresponding to a Nash equilibrium of a game constitutes one's best response to the actions of the other players – provided they are in (the same) Nash equilibrium – playing a strategy that is part of a Nash equilibrium seems appropriate. This normative use of game theory has also come under criticism.

Economics and Business

Game theory is a major method used in mathematical economics and business for modeling competing behaviors of interacting agents. Applications include a wide array of economic phenomena and approaches, such as auctions, bargaining, mergers & acquisitions pricing, fair division, duopolies, oligopolies, social network formation, agent-based computational economics, general equilibrium, mechanism design, and voting systems; and across such broad areas as experimental economics, behavioral economics, information economics, industrial organization, and political economy.

This research usually focuses on particular sets of strategies known as solution concepts or equilibria. A common assumption is that players act rationally. In non-cooperative games, the most famous of these is the Nash equilibrium. A set of strategies is a Nash equilibrium if each represents a best response to the other strategies. If all the players are playing the strategies in a Nash equilibrium, they have no unilateral incentive to deviate, since their strategy is the best they can do given what others are doing.

The payoffs of the game are generally taken to represent the utility of individual players.

A prototypical paper on game theory in economics begins by presenting a game that is an abstraction of a particular economic situation. One or more solution concepts are chosen, and the author demonstrates which strategy sets in the presented game are equilibria of the appropriate type. Naturally one might wonder to what use this information should be put. Economists and business professors suggest two primary uses (noted above): *descriptive* and *prescriptive*.

Political Science

The application of game theory to political science is focused in the overlapping areas of fair division, political economy, public choice, war bargaining, positive political theory, and social choice theory. In each of these areas, researchers have developed game-theoretic models in which the players are often voters, states, special interest groups, and politicians.

Early examples of game theory applied to political science are provided by Anthony Downs. In his book *An Economic Theory of Democracy*, he applies the Hotelling firm location model to the political process. In the Downsian model, political candidates commit to ideologies on a one-dimensional policy space. Downs first shows how the political candidates will converge to the ideology preferred by the median voter if voters are fully informed, but then argues that voters choose to remain rationally ignorant which allows for candidate divergence. Game Theory was applied in 1962 to the Cuban missile crisis during the presidency of John F. Kennedy.

It has also been proposed that game theory explains the stability of any form of political government. Taking the simplest case of a monarchy, for example, the king, being only one person, does not and cannot maintain his authority by personally exercising physical control over all or even any significant number of his subjects. Sovereign control is instead explained by the recognition by each citizen that all other citizens expect each other to view the king (or other established government) as the person whose orders will be followed. Coordinating communication among citizens to replace the sovereign is effectively barred, since conspiracy to replace the sovereign is generally punishable as a crime. Thus, in a process that can be modeled by variants of the prisoner›s dilemma, during periods of stability no citizen will find it rational to move to replace the sovereign, even if all the citizens know they would be better off if they were all to act collectively.

A game-theoretic explanation for democratic peace is that public and open debate in democracies send clear and reliable information regarding their intentions to other states. In contrast, it is difficult to know the intentions of nondemocratic leaders, what effect concessions will have, and if promises will be kept. Thus there will be mistrust and unwillingness to make concessions if at least one of the parties in a dispute is a non-democracy.

Game theory could also help predict a nation›s responses when there is a new rule or law to be applied to that nation. One example would be Peter John Wood›s (2013) research when he looked into what nations could do to help reduce climate change. Wood thought this could be accomplished by making treaties with other nations to reduce green house gas emissions. However, he concluded that this idea could not work because it would create a prisoner›s dilemma to the nations.

Biology

	Hawk	Dove
Hawk	20, 20	80, 40
Dove	40, 80	60, 60

Unlike those in economics, the payoffs for games in biology are often interpreted as corresponding to fitness. In addition, the focus has been less on equilibria that correspond to a notion of rationality and more on ones that would be maintained by evolutionary forces. The best known equilibrium in biology is known as the *evolutionarily stable strategy* (ESS), first introduced in (Smith & Price 1973). Although its initial motivation did not involve any of the mental requirements of the Nash equilibrium, every ESS is a Nash equilibrium.

In biology, game theory has been used as a model to understand many different phenomena. It was first used to explain the evolution (and stability) of the approximate 1:1 sex ratios. (Fisher 1930) suggested that the 1:1 sex ratios are a result of evolutionary forces acting on individuals who could be seen as trying to maximize their number of grandchildren.

Additionally, biologists have used evolutionary game theory and the ESS to explain the emergence of animal communication. The analysis of signaling games and other communication games has provided insight into the evolution of communication among animals. For example, the mobbing behavior of many species, in which a large number of prey animals attack a larger predator, seems to be an example of spontaneous emergent organization. Ants have also been shown to exhibit feed-forward behavior akin to fashion (see Paul Ormerod›s *Butterfly Economics*).

Biologists have used the game of chicken to analyze fighting behavior and territoriality.

According to Maynard Smith, in the preface to *Evolution and the Theory of Games*, "paradoxically, it has turned out that game theory is more readily applied to biology than to the field of economic behaviour for which it was originally designed". Evolutionary game theory has been used to explain many seemingly incongruous phenomena in nature.

One such phenomenon is known as biological altruism. This is a situation in which an organism appears to act in a way that benefits other organisms and is detrimental to itself. This is distinct from traditional notions of altruism because such actions are not conscious, but appear to be evolutionary adaptations to increase overall fitness. Examples can be found in species ranging from vampire bats that regurgitate blood they have obtained from a night's hunting and give it to group members who have failed to feed, to worker bees that care for the queen bee for their entire lives and never mate, to vervet monkeys that warn group members of a predator's approach, even when it endangers that individual's chance of survival. All of these actions increase the overall fitness of a group, but occur at a cost to the individual.

Evolutionary game theory explains this altruism with the idea of kin selection. Altruists discriminate between the individuals they help and favor relatives. Hamilton›s rule explains the evolutionary rationale behind this selection with the equation $c<b*r$ where the cost (c) to the altruist must be less than the benefit (b) to the recipient multiplied by the coefficient of relatedness (r). The more closely related two organisms are causes the incidences of altruism to increase because they share many of the same alleles. This means that the altruistic individual, by ensuring that the alleles of its close relative are passed on, (through survival of its offspring) can forgo the option of having offspring itself because the same number of alleles are passed on. Helping a sibling for example (in diploid animals), has a coefficient of ½, because (on average) an individual shares ½ of the alleles in its sibling's offspring. Ensuring that enough of a sibling's offspring survive to adulthood precludes the necessity of the altruistic individual producing offspring. The coefficient values depend heavily on the scope of the playing field; for example if the choice of whom to favor includes all genetic living things, not just all relatives, we assume the discrepancy between all humans only accounts for approximately 1% of the diversity in the playing field, a co-efficient that was ½ in the smaller field becomes 0.995. Similarly if it is considered that information other than that of a genetic nature (e.g. epigenetics, religion, science, etc.) persisted through time the playing field becomes larger still, and the discrepancies smaller.

Computer Science and Logic

Game theory has come to play an increasingly important role in logic and in computer science. Several logical theories have a basis in game semantics. In addition, computer scientists have used games to model interactive computations. Also, game theory provides a theoretical basis to the field of multi-agent systems.

Separately, game theory has played a role in online algorithms; in particular, the k-server problem, which has in the past been referred to as *games with moving costs* and *request-answer games*. Yao's principle is a game-theoretic technique for proving lower bounds on the computational complexity of randomized algorithms, especially online algorithms.

The emergence of the internet has motivated the development of algorithms for finding equilibria in games, markets, computational auctions, peer-to-peer systems, and security and information markets. Algorithmic game theory and within it algorithmic mechanism design combine computational algorithm design and analysis of complex systems with economic theory.

	Stag	Hare
Stag	3, 3	0, 2
Hare	2, 0	2, 2

Stag hunt

Stag hunt

Philosophy

Game theory has been put to several uses in philosophy. Responding to two papers by W.V.O. Quine (1960, 1967), Lewis (1969) used game theory to develop a philosophical account of convention. In so doing, he provided the first analysis of common knowledge and employed it in analyzing play in coordination games. In addition, he first suggested that one can understand meaning in terms of signaling games. This later suggestion has been pursued by several philosophers since Lewis. Following Lewis (1969) game-theoretic account of conventions, Edna Ullmann-Margalit (1977) and Bicchieri (2006) have developed theories of social norms that define them as Nash equilibria that result from transforming a mixed-motive game into a coordination game.

Game theory has also challenged philosophers to think in terms of interactive epistemology: what it means for a collective to have common beliefs or knowledge, and what are the consequences of this knowledge for the social outcomes resulting from agents› interactions. Philosophers who have worked in this area include Bicchieri (1989, 1993), Skyrms (1990), and Stalnaker (1999).

In ethics, some authors have attempted to pursue Thomas Hobbes› project of deriving morality from self-interest. Since games like the prisoner›s dilemma present an apparent conflict between morality and self-interest, explaining why cooperation is required by self-interest is an important component of this project. This general strategy is a component of the general social contract view in political philosophy (for examples, see Gauthier (1986) and Kavka (1986)).

Other authors have attempted to use evolutionary game theory in order to explain the emergence of human attitudes about morality and corresponding animal behaviors. These authors look at several games including the prisoner's dilemma, stag hunt, and the Nash bargaining game as providing an explanation for the emergence of attitudes about morality (see, e.g., Skyrms (1996, 2004) and Sober and Wilson (1999)).

In popular Culture

Based on the book by Sylvia Nasar, the life story of game theorist and mathematician John Nash was turned into the biopic *A Beautiful Mind* starring Russell Crowe.

Games theory and theory of games are mentioned in the military science fiction novel *Starship Troopers* by Robert A. Heinlein. In the 1997 film of the same name, the character Carl Jenkins refers to his assignment to military intelligence as to games and theory.

The film *Dr. Strangelove* satirizes game theoretic ideas about deterrence theory. For example, nuclear deterrence depends on the threat to retaliate catastrophically if a nuclear attack is detected. A game theorist might argue that such threats can fail to be *credible*, in the sense that they can lead to subgame imperfect equilibria. The movie takes this idea one step further, with the Russians irrevocably committing to a catastrophic nuclear response without making the threat public.

Liar Game is a popular Japanese Manga, television program and movie, where each episode presents the main characters with a Game Theory type game. The show's supporting characters reflect and explore game theory's predictions around self-preservation strategies used in each challenge. The main character however, who is portrayed as an innocent, naive and good hearted young lady Kansaki Nao, always attempts to convince the other players to follow a mutually beneficial strategy where everybody wins. Kansaki Nao›s seemingly simple strategies that appear to be the product of her innocent good nature actually represent optimal equilibrium solutions which Game Theory attempts to solve. Other players however, usually use her naivety against her to follow strategies that serve self-preservation. The show improvises heavily on Game Theory predictions and strategies to provide each episode›s script, the players decisions. In a sense, each episode exhibits a Game Theory game and the strategies/ equilibria/ solutions provide the script which is coloured in by the actors.

REFERENCES

1. Myerson, Roger B. (1991). Game Theory: Analysis of Conflict, Harvard University Press, p. 1. Chapter-preview links, pp. vii–xi.
2. Bellhouse, David (2007), "The Problem of Waldegrave" (PDF), Journal Électronique d›Histoire des Probabilités et de la Statistique, 3 (2)
3. James Madison, Vices of the Political System of the United States, April 1787.
4. Jack Rakove, James Madison and the Constitution, History Now, Issue 13, September 2007. Archived 11 April 2009 at the Wayback Machine.
5. Screpanti; Ernesto; Zamagni; Stefano (2005). An Outline of the History of Economic Thought› (2nd ed.). Oxford University Press.
6. Neumann, J. v. (1928), Zur Theorie der Gesellschaftsspiele, Mathematische Annalen, 100 (1): 295–320, doi:10.1007/BF01448847 English translation: Tucker, A. W.; Luce, R. D., eds. (1959), On the Theory of Games of Strategy, Contributions to the Theory of Games, 4, pp. 13–42
7. Mirowski, Philip (1992). What Were von Neumann and Morgenstern Trying to Accomplish?. In Weintraub, E. Roy. Toward a History of Game Theory. Durham: Duke University Press. pp. 113–147. ISBN 0-8223-1253-0.
8. Leonard, Robert (2010), Von Neumann, Morgenstern, and the Creation of Game Theory, New York: Cambridge University Press, ISBN 9780521562669
9. Prisoner›s Dilemma. Plato.stanford.edu (September 4, 1997). Retrieved on 2013-01-03.
10. Although common knowledge was first discussed by the philosopher David Lewis in his dissertation (and later book) Convention in the late 1960s, it was not widely considered by economists until Robert Aumann›s work in the 1970s.
11. Owen, Guillermo (1995). Game Theory: Third Edition. Bingley: Emerald Group Publishing. p. 11. ISBN 0-12-531151-6.
12. Owen, Guillermo (1995). Game Theory: Third Edition. Bingley: Emerald Group Publishing. p. 4. ISBN 0-12-531151-6.

13. Leyton-Brown & Shoham (2008), p. 60.
14. Jörg Bewersdorff (2005), Luck, logic, and white lies: the mathematics of games, A K Peters, Ltd., pp. ix–xii and chapter 31, ISBN 978-1-56881-210-6
15. Albert, Michael H.; Nowakowski, Richard J.; Wolfe, David (2007), Lessons in Play: In Introduction to Combinatorial Game Theory, A K Peters Ltd, pp. 3–4, ISBN 978-1-56881-277-9
16. Beck, József (2008), Combinatorial games: tic-tac-toe theory, Cambridge University Press, pp. 1–3, ISBN 978-0-521-46100-9
17. Robert A. Hearn; Erik D. Demaine (2009), Games, Puzzles, and Computation, A K Peters, Ltd., ISBN 978-1-56881-322-6
18. M. Tim Jones (2008), Artificial Intelligence: A Systems Approach, Jones & Bartlett Learning, pp. 106–118, ISBN 978-0-7637-7337-3
19. (Russian) Petrosjan, L.A. and Murzov, N.V. (1966). Game-theoretic problems of mechanics. Litovsk. Mat. Sb. 6, 423–433.
20. Luce & Raiffa (1957).
21. Webb (2007).
22. Osborne & Rubinstein (1994).
23. Hugh Brendan McMahan (2006), Robust Planning in Domains with Stochastic Outcomes, Adversaries, and Partial Observability, CMU-CS-06-166, pp. 3–4
24. Howard (1971).
25. Wenliang Wang (2015). Pooling Game Theory and Public Pension Plan. ISBN 978-1507658246.
26. Eric Rasmusen (2007). Games and Information, 4th ed. Description and chapter-preview
27. David M. Kreps (1990). Game Theory and Economic Modelling. Description.
28. R. Aumann and S. Hart, ed. (1992, 2002). Handbook of Game Theory with Economic Applications v. 1, ch. 3–6 and v. 3, ch. 43.
29. Fudenberg & Tirole (1991), p. 67.
30. Williams, Paul D. (2013). Security Studies: an Introduction (second edition). 2 Park Square, Milton Park, Abingdon, Oxon, OX14 4RN: Routledge. pp. 55–56.
31. Leyton-Brown & Shoham (2008), p. 35.
32. Colin F. Camerer (2003). Behavioral Game Theory: Experiments in Strategic Interaction, pp. 5–7 (scroll to at 1.1 What Is Game Theory Good For?).
33. Ross, Don. Game Theory. The Stanford Encyclopedia of Philosophy (Spring 2008 Edition). Edward N. Zalta (ed.). Retrieved 2008-08-21.
34. Experimental work in game theory goes by many names, experimental economics, behavioral economics, and behavioural game theory are several. For a recent discussion, see Colin F. Camerer (2003). Behavioral Game Theory: Experiments in Strategic Interaction (description and Introduction, pp. 1–25).
35. At JEL:C7 of the Journal of Economic Literature classification codes.

36. R.J. Aumann (2008). game theory, The New Palgrave Dictionary of Economics, 2nd Edition. Abstract. Archived 15 May 2011 at the Wayback Machine.
37. Martin Shubik (1981). Game Theory Models and Methods in Political Economy, in Kenneth Arrow and Michael Intriligator, ed., Handbook of Mathematical Economics, v. 1, pp. 285–330 doi:10.1016/S1573-4382(81)01011-4.
38. Carl Shapiro (1989). The Theory of Business Strategy, RAND Journal of Economics, 20(1), pp. 125–137 JSTOR 2555656.
39. N. Agarwal and P. Zeephongsekul. Psychological Pricing in Mergers & Acquisitions using Game Theory, School of Mathematics and Geospatial Sciences, RMIT University, Melbourne
40. Leigh Tesfatsion (2006). Agent-Based Computational Economics: A Constructive Approach to Economic Theory, ch. 16, Handbook of Computational Economics, v. 2, pp. 831-880 doi:10.1016/S1574-0021(05)02016-2.
41. Joseph Y. Halpern (2008). computer science and game theory, The New Palgrave Dictionary of Economics, 2nd Edition. Abstract.
42. From The New Palgrave Dictionary of Economics (2008), 2nd Edition:
43. Roger B. Myerson. mechanism design. Abstract Archived 23 November 2011 at the Wayback Machine..
44. _____. revelation principle. Abstract.
45. Tuomas Sandholm. computing in mechanism design. Abstract. Archived 23 November 2011 at the Wayback Machine.
46. Noam Nisan and Amir Ronen (2001). Algorithmic Mechanism Design, Games and Economic Behavior, 35(1–2), pp. 166–196.
47. Noam Nisan et al., ed. (2007). Algorithmic Game Theory, Cambridge University Press. Description. Archived 23 November 2011 at the Wayback Machine.
48. Aumann, R. and Hart, S. (eds.) (1994). Handbook of Game Theory with Economic Applications, v. 2, ch. 30: Voting Procedures and ch. 31: Social Choice.
49. Vernon L. Smith, 1992. Game Theory and Experimental Economics: Beginnings and Early Influences, in E. R. Weintraub, ed., Towards a History of Game Theory, pp. 241–282.
50. _____, 2001. Experimental Economics, International Encyclopedia of the Social & Behavioral Sciences, pp. 5100–5108. Abstract per sect. 1.1 & 2.1.
51. Charles R. Plott and Vernon L. Smith, ed., 2008. Handbook of Experimental Economics Results, v. 1, Elsevier, Part 4, Games, ch. 45–66.
52. Vincent P. Crawford (1997). Theory and Experiment in the Analysis of Strategic Interaction, in Advances in Economics and Econometrics: Theory and Applications, pp. 206–242. Cambridge. Reprinted in Colin F. Camerer et al., ed. (2003). Advances in Behavioral Economics, Princeton. 1986–2003 papers. Description, preview, Princeton, ch. 12.
53. Martin Shubik, 2002. Game Theory and Experimental Gaming, in R. Aumann and S. Hart, ed., Handbook of Game Theory with Economic Applications,

Elsevier, v. 3, pp. 2327–2351. doi:10.1016/S1574-0005(02)03025-4.

54. From The New Palgrave Dictionary of Economics (2008), 2nd Edition:
55. Faruk Gul. behavioural economics and game theory. Abstract.
56. Colin F. Camerer. behavioral game theory. Abstract. Archived 23 November 2011 at the Wayback Machine.
57. _____ (1997). Progress in Behavioral Game Theory, Journal of Economic Perspectives, 11(4), p. 172, pp. 167–188.
58. _____ (2003). Behavioral Game Theory, Princeton. Description, preview ([ctrl]+), and ch. 1 link.
59. _____, George Loewenstein, and Matthew Rabin, ed. (2003). Advances in Behavioral Economics, Princeton. 1986–2003 papers. Description, contents, and preview.
60. Drew Fudenberg (2006). Advancing Beyond Advances in Behavioral Economics, Journal of Economic Literature, 44(3), pp. 694–711 JSTOR 30032349.
61. Jean Tirole (1988). The Theory of Industrial Organization, MIT Press. Description and chapter-preview links, pp. vii–ix, General Organization, pp. 5–6, and Non-Cooperative Game Theory: A User›s Guide Manual,› ch. 11, pp. 423–59.
62. Kyle Bagwell and Asher Wolinsky (2002). Game theory and Industrial Organization, ch. 49, Handbook of Game Theory with Economic Applications, v. 3, pp. 1851–1895.
63. Martin Shubik (1959). Strategy and Market Structure: Competition, Oligopoly, and the Theory of Games, Wiley. Description and review extract.
64. _____ with Richard Levitan (1980). Market Structure and Behavior, Harvard University Press. Review extract. Archived 15 March 2010 at the Wayback Machine.
65. Martin Shubik (1981). Game Theory Models and Methods in Political Economy, in Handbook of Mathematical Economics, v. 1, pp. 285–330 doi:10.1016/S1573-4382(81)01011-4.
66. •_____ (1987). A Game-Theoretic Approach to Political Economy. MIT Press. Description. Archived 29 June 2011 at the Wayback Machine.
67. Martin Shubik (1978). Game Theory: Economic Applications, in W. Kruskal and J.M. Tanur, ed., International Encyclopedia of Statistics, v. 2, pp. 372–78.
68. Robert Aumann and Sergiu Hart, ed. Handbook of Game Theory with Economic Applications (scrollable to chapter-outline or abstract links):
69. 1992. v. 1; 1994. v. 2; 2002. v. 3.
70. Game-theoretic model to examine the two tradeoffs in the acquisition of information for a careful balancing act Research paper INSEAD
71. Options Games: Balancing the trade-off between flexibility and commitment Archived 20 June 2013 at the Wayback Machine.. Europeanfinancialreview.com (2012-02-15). Retrieved on 2013-01-03. Archived 20 June 2013 at the Wayback Machine.

72. (Downs 1957)
73. Steven J. Brams, Game theory and the Cuban missile crisis, Plus Magazine, 1 January 2001, accessed 31 January 2016.
74. Morrison, A. S. (2013). Yes, Law is the Command of the Sovereign. doi:10.2139/ssrn.2371076.
75. Levy, G.; Razin, R. (2004). It Takes Two: An Explanation for the Democratic Peace. Journal of the European Economic Association. 2: 1. doi:10.1162/154247604323015463. JSTOR 40004867.
76. Wood, Peter John (2011). Climate change and game theory. Ecological Economics Review. 1219: 153–70. doi:10.1111/j.1749-6632.2010.05891.x.
77. Harper & Maynard Smith (2003).
78. Maynard Smith, J. (1974). The theory of games and the evolution of animal conflicts. Journal of Theoretical Biology. 47 (1): 209–221. doi:10.1016/0022-5193(74)90110-6. PMID 4459582.
79. Evolutionary Game Theory (Stanford Encyclopedia of Philosophy). Plato.stanford.edu. Retrieved on 2013-01-03.
80. Biological Altruism (Stanford Encyclopedia of Philosophy). Seop.leeds.ac.uk. Retrieved on 2013-01-03.
81. (Ben David, Borodin & Karp et al. 1994)
82. Noam Nisan et al., ed. (2007). Algorithmic Game Theory, Cambridge University Press. Description. Archived 5 May 2012 at the Wayback Machine.
83. Nisan, Noam; Ronen, Amir (2001), Algorithmic Mechanism Design (PDF), Games and Economic Behavior, 35 (1–2): 166–196, doi:10.1006/game.1999.0790
84. Joseph Y. Halpern (2008). computer science and game theory, The New Palgrave Dictionary of Economics, 2nd Edition. Abstract.
85. Shoham, Yoav (2008), Computer Science and Game Theory (PDF), Communications of the ACM, 51 (8): 75–79, doi:10.1145/1378704.1378721
86. Littman, Amy; Littman, Michael L. (2007), Introduction to the Special Issue on Learning and Computational Game Theory, Machine Learning, 67 (1–2): 3–6, doi:10.1007/s10994-007-0770-1
87. (Skyrms (1996), Grim, Kokalis, and Alai-Tafti et al. (2004)).
88. Ullmann-Margalit, E. (1977), The Emergence of Norms, Oxford University Press, ISBN 0198244118
89. Bicchieri, C. (2006), The Grammar of Society: the Nature and Dynamics of Social Norms, Cambridge University Press, ISBN 0521573726
90. Bicchieri, Cristina (1989), Self-Refuting Theories of Strategic Interaction: A Paradox of Common Knowledge, Erkenntnis, 30 (1–2): 69–85, doi:10.1007/BF00184816
91. Bicchieri, Cristina (1993), Rationality and Coordination, Cambridge University Press, ISBN 0-521-57444-7

92. The Dynamics of Rational Deliberation, Harvard University Press, 1990, ISBN 067421885X
93. Bicchieri, Cristina; Jeffrey, Richard; Skyrms, Brian, eds. (1999), Knowledge, Belief, and Counterfactual Reasoning in Games, The Logic of Strategy, New York: Oxford University Press, ISBN 0195117158
94. For a more detailed discussion of the use of game theory in ethics, see the Stanford Encyclopedia of Philosophy›s entry game theory and ethics.
95. Nasar, Sylvia (1998) A Beautiful Mind, Simon & Schuster. ISBN 0-684-81906-6.
96. Singh, Simon (June 14, 1998) Between Genius and Madness, New York Times.
97. Heinlein, Robert A. (1959), Starship Troopers

Index

A

Analytic Combinatorics 86
Axiomatic Set Theory 39
Axiom of determinacy (AD) 42

B

Basic Concepts And Notation 37

C

Cartesian product 38, 54, 57, 58, 64, 66, 67, 77
central limit theorem (CLT) 120
Combinatorial game theory (CGT) 93
combinatorics 83, 84, 85, 86, 87, 88, 89, 90, 91, 92, 93

D

Differentially closed fields (DCF) 22

F

Fuzzy Set Theory 41

G

Generalized Central Limit Theorem (GCLT) 120

H

hereditary 108
hierarchy 39, 40, 41
hierarchy. 39, 40, 41

M

MorenoSociogram 104
Morse–Kelley set theory (MK) 9

N

Neumann–Bernays–Gödel set theory (NBG) 9
Neumann universe 38, 39

O

Ontology 38
Ostomachion 85

P

permutation 83

R

Relational model (RM) 51
RVA – Relation-valued attribute 58

Z

Zermelo–Fraenkel set theory (ZF) 5, 9
Zermelo–Fraenkel set theory (ZFC) 39